D1247857

Securing the Safe Performance of Graphite Reactor Cores

Securing the Safe Performance of Graphite Reactor Cores

Edited by

Gareth B. Neighbour
Department of Engineering, University of Hull, Hull, UK

RSCPublishing

The proceedings of the meeting on Securing the Safe Performance of Graphite Reactor Cores held 24th-26th November 2008 at the University of Nottingham and run under the auspices of the British Carbon Group.

Special Publication No. 328

ISBN: 978-1-84755-913-5

A catalogue record for this book is available from the British Library

Published by The Royal Society of Chemistry,
Thomas Graham House, Science Park, Milton Road,
Cambridge CB4 0WF, UK

Registered Charity Number 207890

For further information see our web site at www.rsc.org

Foreword

All graphite moderated reactors suffer ageing and degradation to the graphite during service. The degradation poses a threat to the functionality of the graphite core, and potentially, the safe operation of the reactor. In particular for AGRs and Magnox reactors in the UK, the concern is with brick cracking (and subsequent change in core behaviour) and graphite weight loss.

The safety requirements for the AGRs can be expressed as:

- Unhindered movement of control rods.
- Adequate cooling of fuel and core.
- Unhindered fuel charge and discharge.

These requirements are assessed in safety cases, and these cases are 'constructed' around five principal legs:

- Core and Component Condition Assessment (CCCA) – graphite properties, weight loss, and brick cracking.
- Damage Tolerance Assessment (DTA) – consequence of cracking on gas flow (and hence fuel temperature), and core distortion.
- Core Monitoring – information collected during reactor operation.
- Core Inspection – information collected during reactor off-load.
- Consequences Assessment and ALARP – lack of cliff edge in consequences, and overall safety envelope.

The vast range of work presented in this book falls into one of these five legs, and the absolute and relative strength of these legs contribute to the overall safety case. The safety cases are now written around Claims, Arguments and Evidence. For example, a Claim is:

> *The condition of the graphite core is essentially intact, that is, the keying system is functional, which is fundamental for the assessment of core distortion. Bricks with bore cracks that affect the keyways are very few in number, there are no keyway root cracked bricks.*

The Arguments are:

- Bore cracking will not threaten the functionality of the keying system. The number of bricks with bore cracks is low and there is evidence that the rate of cracking is reducing and could possibly have stopped altogether. Very few of the bore cracks are axial cracks, and it is only axial cracks that could affect the keying system.

- The reactor cores will therefore remain essentially intact, with a functional keying system, at least until the onset of keyway root cracking.

- The onset of keyway root cracking (calculated on a conservative basis) is not predicted before 26.2 reactor full-power-years (fpy). Even if full power operation is assumed, this core irradiation will not be reached before 2012, even for the lead reactor.

- Operation at reduced power (between ~60% and 100%) will delay the onset of keyway root cracking, as the equivalent core irradiation will be achieved at a later date.

- A re-assessment of material properties is underway, and the re-assessed properties are more representative of behaviour in the cores since they have been determined including the AGR trepanned sample data where the previous assessments (which gave 26.2 reactor fpy) used Material Test Reactor data alone. As an examination of the conservatism in the analysis, subsequent scoping brick stress analysis indicates that the onset of keyway root cracking would be later than the current calculations, by at least 3 reactor fpy.

And finally the Evidence ... well a number of examples will be presented during this book!

Dr James Reed
British Energy Part of EDF
March 2010

Preface

The United Kingdom has a number of gas-cooled, graphite-moderated nuclear reactors that are either nearing the end of their generating life, seeking life extension, or in the latter period of their design lives. As the graphite components of the reactor cores cannot be replaced, it is important to ensure that effective strategies are in place to secure safe and reliable operation ultimately beyond their planned design life. The importance of extending the safe operating life of a reactor is increasing due to the political and socio-economic demands to reduce greenhouses gases and to diversify energy supply.

This book represents the record and output from the 'Securing the Safe Performance of Graphite Reactor Cores' conference which took place between the 24th and 26th November 2008 at the University of Nottingham. The conference was very well attended and was organised with support from all the major stakeholders in the UK, namely AMEC, Atkins, British Energy, Frazer Nash Consultants, Health and Sagety Executive, Magnox North, National Nuclear Laboratory, NRG, and Serco. This conference on securing the safe performance of graphite reactor cores represented a second gathering of the most eminent specialists in the field from around the world. The first was three years previously at the very successful 'Management of Ageing Processes in Graphite Reactor Cores' meeting at the University of Cardiff in November 2005.

The issues associated with polygranular graphite as a moderator have remained in the forefront of reactor technology since the creation of the first reactor pile, CP1, in Chicago in 1942. Whilst the amazing story of graphite moderators, in terms of the British experience, was reviewed during the first meeting, the second conference very much looks to the future to secure the continued safe and efficient operation of the remaining UK graphite moderated reactors. I believe the conference was very successful, and thus, this book provides an opportunity to review the recent advances in knowledge and understanding which will remain significant in years to come and very relevant to the designers and operators of Generation III+ and IV graphite-moderated reactors. Further, this book seeks to review recent developments in the methodologies that are being used to assess the condition of the graphite and to consider some related subjects such as plant modifications for safe operation, graphite decommissioning and the design of future reactor systems. That is, this book aims to be forward facing towards the challenges ahead, particularly in relation to the following areas:

- Assessing Core Behaviour and Methodologies
- Surveillance and Test Methods
- Prediction of Component Performance
- Plant Performance
- Graphite Reactor Decommissioning
- Lessons Learned and Forward Strategy

Gareth B Neighbour
University of Hull
March 2010

Acknowledgements

The editor would like to express his heartfelt thanks to the many authors who contributed to this volume. He would also like to thank his wife Angela and children, Caitlin and Rhiannon, for their love, support and tolerance in preparing this book. Finally, but not least, the editor would like to recognise the support of the British Carbon Group (registered charity 207890 which itself is affiliated to The Royal Society of Chemistry, The Institute of Physics and The Society of Chemical Industry) and the Organising Committee of the conference without which this book would not exist.

Organising Committee

Prof Brian McEnaney	*University of Bath & Conference Chair*
Dr James Reed	*British Energy*
Prof Barry Marsden	*University of Manchester*
Dr Gareth Neighbour (Editor)	*The University of Hull &*
	Chairman, British Carbon Group
Dr Anthony Wickham (Secretary)	*Conference Manager*
Dr Catherine Earl	*Atkins*

On behalf of the British Carbon Group, the editor would also like to express thanks to the following organisation for financial support for the conference which led to this book.

- British Energy
- AMEC
- Atkins
- Frazer Nash Consultants
- Health and Safety Executive
- Magnox North
- National Nuclear Laboratory
- NRG
- Serco

Contents

Part D - Prediction of Component Performance

Part E - Plant Performance

Part F - Graphite Reactor Decommissioning

Part A

General Introduction

A Look into the Past – How it All Began

Derek Dominey[1]

Independent Consultant, Gloucester, UK
Email: derek.dominey@ukgateway.net

Abstract
The paper records the plenary lecture given by Dr Derek Dominey. The paper will recall his experience of the early days of the research which underwrote the design and operation of the graphite moderated commercial reactors in the UK. In 1959, he joined a team at Harwell involved in the research into the radiation induced carbon dioxide-graphite reaction. His role was to pioneer the use of Carbon-14 which he had used in his research at Oxford. BEPO and later DIDO were used as radiation sources. In 1966, the research was continued at Berkeley Nuclear Laboratories using a large Co-60 source. The graphite monitoring programme for the AGRs was developed at Berkeley Nuclear Laboratories (BNL) during this period. His paper will describe the experimental methods and some of the personalities involved.

Keywords
BEPO, DIDO, Berkeley Nuclear Labs, Radiolytic oxidation

PLENARY LECTURE

I am very conscious of the honour done to me by the request from the organising committee to give this plenary lecture. I have been involved in the chemistry of graphite, on and off, for over 50 years and this is, perhaps, an appropriate time to look back at where it all began.

I have taken some inspiration in preparing the talk from two books. The first by Professor Otto Frisch is entitled "What Little I Remember" and in the Preface he acknowledged that his memory was not always accurate and that he had tried to bring to life some of the people he had met. The second was by Spike Milligan entitled "Adolph Hitler; My Part in his Downfall". So my story relies on an incomplete and maybe inaccurate memory and I emphasise that personally I played a very small part in what has been a remarkable success story. I will refer to the part played by many individuals, but I could mention so many more and I apologise to those who I have neglected to name.

As an aside, it is interesting to note that in 1975 Otto Frisch wrote a spoof article entitled "Are coal-driven power stations feasible" supposedly written at a time when uranium supplies were becoming exhausted and a new source of energy was needed. The article concluded that the problems with safety would rule out the use of coal. The noxious gases produced by combustion would have *"to be collected in suitable containers, pending chemical detoxification. Alternatively the waste might be mixed with hydrogen and filled into large balloons, which are subsequently released"*. This before any talk of global warming.

---

[1] Independent Consultant, formally Technical Support Director - Nuclear Electric, Chairman of the Nuclear Safety Committee, Director of Plant Engineering (Nuclear) in Operational Engineering Division of CEGB, Director of Resource Planning South East Region CEGB, Scientific Services Controller South East Region CEGB, Head of Materials Science Division at Berkeley Nuclear Lab.

So, how did it all begin for me? I was at school when the atomic bombs were dropped on Japan. Post war, the hot topic was how this new form of energy would be turned to practical use. There was a series of conferences in Geneva on "The Peaceful Uses of Atomic Energy". The proceedings still make interesting reading. I was intrigued and felt that this exciting prospect would play a part in my subsequent career. And so it proved.

I read Chemistry at Oxford. This was a four year course, the last of which involved a full time research project. I persuaded Professor Hinshelwood to take me as a member of his team because he was the only person supervising work on radioactive isotopes. He was investigating the metabolism of bacteria by measuring the uptake of labelled nutrients labelled with Sulphur-35, Phosphorus-32 or Carbon-14. During this period I also worked on isotope effects, measuring the different rates of reactions with compounds containing Carbon-12 and Carbon-13 (which forms about 1% of natural carbon). Measurements were made by mass spectrometry. This work was done in collaboration with Professor Gus A. Ropp from Tennessee and my first published paper was in the Journal of the American Chemical Society in 1957. Gus was a very tall, laconic guy who became a great friend. He and his wife explored the British Isles mainly at weekends in a tiny Volkswagon which he bought in Germany and subsequently took home to the USA. His other hobby was photography, but he took so long to get an accurate reading on his light meter that the light had usually changed by the time he clicked the shutter!

Subsequently, I joined the team which followed up the Professor's earlier studies on the mechanism of hydrocarbon pyrolysis with Carbon-14 labelled compounds. All the compounds we used had to be synthesised. We obtained our supplies of radioactive isotopes, which were produced at Harwell in BEPO, from the Radio Chemical Centre at Amersham who could supply any form of Carbon-14 provided it was CO_2! We also had to develop methods for measuring the very low energy beta radiation from Carbon-14. The CO_2 was converted to Barium Carbonate which was filtered and dried and then counted with a thin window Geiger-Muller counter. Corrections had to be applied for the thickness of the sample because the radiation was absorbed by the sample itself. To measure the activity of any other compound, it had to be converted to CO_2 and processed in the same way. This was a tedious process and eventually we developed a windowless scintillation counter on the end of a gas chromatograph, packing the columns ourselves, and we were able to measure the activity of individual gases. Our sole apparatus for working out our results was a mechanical calculator which involved much handle turning; it was shared between a team of 5 which included a Canadian, an Indian, two South Africans and myself and we became adept at managing international disputes! Sometimes we worked all night on the basis that as long as the equipment was working - keep going. So CO_2 and radiation entered my life.

After graduation, I applied for a Research Fellowship at Harwell as the next step towards my goal. After months of investigation by the security agencies during which I turned down several other opportunities, including one to join a team developing a reactor to power aircraft, I was allowed to take up the Fellowship. Remember this was not long after the unmasking of Klaus Fuchs in 1950 so the caution was understandable.

I arrived at Harwell in 1959 in the middle of a Geneva Conference and met the Head of Chemistry Division, Bob Spence, who gave me a month to look round the laboratories and go back to him with a proposal for research when he returned from Geneva. I duly did this and as a result joined the Pile Irradiation Group under John Wright.

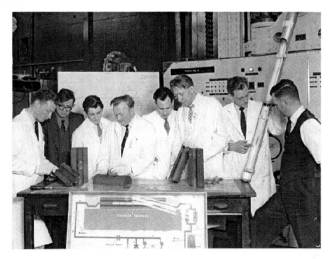

FIGURE 1: John Wright with members of the Pile Radiation Chemistry Group (L to R - Fred Walker, Barry Copestake (GEC), Tony Poole, Bill Marsh, Frank Feates, Mike Tomlinson, Roger Sach and John Wright).

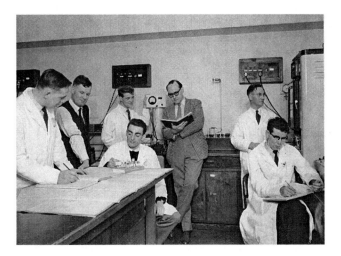

FIGURE 2: John Wright with members of the Pile Radiation Chemistry Group (L to R – Bob Waite, John Wright, Alan Morris, Norman Taylor, Ken Linacre, Ron Thomas and Taffy Davies).

I was fortunate in starting my career under John and I learned an enormous amount from him and the members of his team. John was methodical and logical in his approach. He had spent time at Chalk River before moving to Harwell. John was a former graduate of my College and he took me under his wing. His criticism and correction of my reports was ruthless and those who subsequently suffered my treatment of their efforts owe much of this to him. He taught us to be rigorous. When he retired John produced several superb reports which summarised the research at that point. The major report (Wright, 1981) contains 373 references and the list reads like a roll of honour of those who did the pioneer work. Sadly

John contracted Alzheimer's disease from which he died. Ironically this disease has played a major role in my life since I retired from full time work in 1995 to look after my wife.

Two other senior members of the group were Mike Tomlinson and Ken Linacre. Mike is a lanky Yorkshireman whose main hobby was climbing. He and his wife, Helen, who worked in the translation service in the Harwell library, would set off for the Scottish mountains on Friday evenings and be back at work on Monday morning. There were no motorways then. On one such trip Mike broke his leg. He was a neighbour of ours and I drove him into work in his Ford Zephyr with his leg propped on the dashboard. When their son was born he was carried up mountains on Mike's back. Mike and Helen eventually emigrated to Canada and Mike played a prominent role in developing Whiteshell at Pinawa. They are still living there in retirement. Ken Linacre was largely responsible for developing calorimetry as the method for measuring absorbed radiation dose. This was essential to underwrite the corrosion measurements which the group made.

So what did I do? At that time the first Magnox reactors at Berkeley and Bradwell were being built. They began operation in 1962. Reactors with higher power were being designed, including the later Magnox reactors and the Advanced Gas-cooled Reactors. Let me remind you of the problem the designers were facing. Carbon dioxide was the chosen coolant partly because of its apparent radiation stability, but it reacted with graphite to produce carbon monoxide. So the moderator / structure of the core would slowly erode. But how slowly? The designers had in mind at least 20 years life and the maximum tolerable weight loss was thought to be about 20%. How do you measure a corrosion rate of 1% per annum with sufficient accuracy to design with confidence a structure that will operate safely for 20 years or more.

The first methods involved measuring the rate of CO production from small graphite samples under irradiation in CO_2 in sealed tubes. By varying the size and shape of the graphite samples in the tubes, the first estimates were made about the lifetime and range of the active species responsible for the corrosion. Use of these data required very large extrapolation and the assumption that the mechanism of the reaction was understood. This programme was managed by staff from GEC at Wembley and the results were published in a series of GEC reports between 1959 and 1962. Norman Corney and Barry Copestake took turns in joining our group to supervise the irradiation of the specimens in BEPO. The analysis was carried out at Wembley. Both were enthusiastic and helpful colleagues. Sadly, I believe that Barry was later killed in the East Midlands air crash.

BEPO was an air cooled graphite moderated research reactor producing about 5MW. There were 3 ways of loading samples into the reactor. Specimens could be loaded into carriers which were put into the reactor during shutdowns; or they could be loaded at power by two devices called the 'Rabbit' and the 'Jumping Jack'.

The 'Rabbit' was a pneumatic tube, like a Lampson tube that I remember being used in some shops to send money from the counter to the cash office and return change. The sample for irradiation was contained in a silica tube with a thin membrane at one end so that the sample tube could be opened in a vacuum system to retrieve and analyse the gas. The silica tube was packed with silica wool inside a plastic container about 3" long and 1" diameter. This was fired into the reactor from the active laboratory. At the end of the experiment, the sample was retrieved the same way. You will appreciate that the silica tube suffered big shocks during this process and the probability of shattering the tube was substantial. The membrane strength was another problem. It had to be strong enough to withstand the impacts, but weak enough to break in the vacuum apparatus. So we always retrieved our samples, at the end of long tongs, with some trepidation. Because the samples had to be transferred to lead pots to allow short lived radiation to decay it was often some time before the tubes could be transferred to the analytical equipment. If at that stage, the tube was found to be damaged or if the membrane could not be broken the experiment was a write off. A story is told in a recent book by Nick Hance (2006) of an occasion when a sample became stuck in the rabbit. Unknown to the operator a section of pipe above ground had been disconnected and a worker's dust cap stuffed into the pipe. The compressed air pressure was wound up to dislodge the rabbit with the result that it was fired through the hangar window like a bullet.

The 'Jumping Jack' was also a pneumatic device, but more gentle on the samples. It was operated from the pile cap and consisted of tubes leading into the active zone which could be accessed separately by rotating a plug by hand. A compressed air line fed in to the bottom of the rig and the gas pressure was adjusted manually by means of a valve at the top. The sample was loaded into a magazine in the plug. The air pressure was adjusted in the loading tube; the pressure was indicated on a dial. The top plug was then rotated so that the sample floated on the air cushion. The pressure was then manually reduced allowing the sample to drift to the bottom of the rig. After the required radiation period, the sample was retrieved by the reverse of this process. This activity required some degree of skill and great concentration. Things did not always go according to plan. One had to be careful to ensure that the compressed air flow was in the correct tube. On one occasion, attempting to retrieve the sample, I inadvertently directed the air flow up the tube that was empty, with the result that I got a blast of air contaminated with radiation in the face. This necessitated a visit to the Medical Department for a shower and some tests to determine how much I had ingested. Fortunately, it was negligible, but I made sure it never happened again.

The process required intense concentration. The shutdown mechanism of BEPO involved driving in control rods with compressed air. They were tested regularly and a warning was always sounded to alert those on the pile cap of the impending big bang. Even then it was a shock when it happened. Imagine then the day when I was crouched over the jumping jack in the course of lowering a sample into the reactor when it tripped. The shock was bad enough, but my worst concern was that I had caused the trip and that I would be blamed by the operators. Fortunately the trip had nothing to do with me.

Before leaving BEPO, I would like to relate the occasion when, in the aftermath of the fire at Windscale, it was decided to release the Wigner energy stored in the BEPO graphite by raising the temperature by a controlled procedure. Thermocouples located throughout the core were connected to an array of chart recorders set around the hanger floor. Each recorder monitored about 10 thermocouples and there was an alarm level on the chart. We observed the rising temperatures as additional heat was supplied and as the Wigner energy was released the slope of the curves increased. This was all done very slowly and took many hours. Eventually the temperature readings levelled off well short of alarm level to everyone's relief.

The measurement of the rate of corrosion by the assay of CO production was laborious and not very sensitive and I was given the task of developing a method of measurement using Carbon-14 employing my earlier experience. With the aid of colleagues from Chemical Engineering Division rods of PGA graphite were labelled with C-14 by induction heating to 2000 °C in the presence of labelled CO followed by an anneal at 2200 °C. Later samples were produced by cracking labelled methane onto an impermeable high density carbon and by heating graphite impregnated with labelled glucose to 1000 °C. These samples were irradiated in a CO_2 atmosphere, initially in sealed tubes. This technique was extensively used in the in-reactor rigs which were developed later.

The other objective of my early research was to investigate why CO_2 was apparently stable under irradiation. Clearly the energy of the radiation was more than sufficient to break the bonds of the molecule so there must be a mechanism for recombination of the initial products of decomposition. The method chosen to investigate this mechanism was to study the exchange of C-14 between CO_2 and labelled CO. The original experiments were done by Don Stranks during a short secondment to Harwell from Leeds University. He coated silica tubes with boron oxide and took advantage of the B(n,alpha)Li reaction to increase the dose rate during irradiation in BEPO by 1 or 2 orders of magnitude. However, the quality of the radiation was quite different from that in a power reactor and we could not be sure that the results were applicable and I did not pursue this line for long.

The assumption made was that CO_2 would initially be decomposed to a CO molecule and an oxygen atom which would then recombine with the labelled CO to form labelled CO_2 the rate of formation of which would be equivalent to the initial rate of decomposition. Although the mechanism was later shown to be much more complex than this simple model these experiments did give early indication of the G value for CO_2 decomposition. Later Frank Palmer joined the group and we collaborated in developing this work which was presented at the first discussion of the Faraday Society which took place outside the UK in The University of Notre Dame near Chicago (Dominey & Palmer, 1963). This Faraday Discussion was chaired by Fred Dainton who decided on the way across the Atlantic that our paper should open the proceedings which meant that I could enjoy the rest of the papers. This technique of using various isotopes, including C-13 and O-18 proved a powerful technique for unravelling the radiation chemistry of CO_2. The work was carried on at Berkeley Nuclear Laboratories by Tony Wickham who gained his PhD for this research (Dominey & Wickham, 1971). His external examiner was Fred Dainton who by that time was Professor of Physical Chemistry at Oxford.

A simultaneous study of the radiation chemistry of CO_2 mixed with various scavengers was carried out in Colin Amphlett's group at Harwell using BEPO and an irradiated fuel-rod assembly (the TIG Pond) at Wantage (Anderson *et al.*, 1962). A joint review of the literature was published with Ron Anderson (Anderson & Dominey, 1968).

Following the Faraday Discussion I was able to visit Oak Ridge National Laboratory and met Samuel Colville Lind (1879-1965) who was one of the pioneers of radiation chemistry. He used alpha radiation to study the radiation chemistry of gases including CO_2 (Lind & Bardwell, 1925). He spent a year working with Marie Curie in Paris and at the Radium Institute in Vienna. In 1913 he joined the US Bureau of Mines Laboratory in Denver where he worked at extracting radium from Carnotite, a uranium ore (potassium uranium vanadate mineral). Lind's close colleague at ORNL, Dr Philip Rudolph, later gave me an extract of a memoir written by Lind about that time which contains some hair-raising descriptions of the process which would certainly not have been allowed within any current safety regulations. He must have ingested a quantity of radioactive material and did suffer radiation burns. However, he was still visiting the laboratory in 1963 and seemed very vigorous. His other passion was fishing. He met his death at the age of 86 while trout fishing in the Clinch River below the Norris Dam. Being deaf, he did not hear the klaxons warning of the opening of the sluices and was drowned. So he died with his boots on.

I also visited the Argonne National Laboratory and was given an Anniversary Issue of the Laboratory Bulletin dated December 1962 which celebrated the 20th anniversary of the first self sustained nuclear chain reaction in the CP-1 reactor. This bulletin gives a first hand account of the start up. One extract will indicate the tension:

> *At 11:35 the automatic safety rod was withdrawn. The control rod was adjusted and the Zip [the emergency rod] was withdrawn. Up went the counters, clicking faster and faster. The graph pen started to climb. Tensely the little group watched and waited entranced by the climbing needle. BANG. As if by a thunder clap the spell was broken. Every man froze and then breathed a sigh of relief when he realised that the automatic rod had slammed home. The safety point at which the rod operated automatically had been set too low. 'I'm hungry' said Fermi, 'Let's go to lunch'.*

The early studies of the CO_2 graphite reaction showed that the rate of corrosion in the new designs of Advanced Gas-cooled Reactors would be excessive and a search for inhibitors began. At Harwell this work was carried out mainly in DIDO which began operation in 1956, opened by Professor Hinshelwood. The rigs were run by Frank Feates and his team and latterly by my team of Harry Morley and Bob Waite. Carbon-14 labelled graphite samples were sealed in stainless steel capsules and gas mixtures passed through them. The compositions of inlet and outlet gases were analysed by gas chromatographs and activity was measured by scintillation counters. We had no computers. Everything was captured on chart recorders which were taken home and measured on the kitchen table. A reactor trip was the opportunity to measure a transient which gave information on relative reaction rates. Unfortunately, the ink or the paper roll sometimes ran out at crucial points so we always tried to be there when anything interesting was happening.

The work at Harwell on the inhibition by methane was coincident with the work in the Windscale AGR by the Group led by Bob Lind at Culcheth. The two groups worked closely together though there was a degree of rivalry which was stimulating and I appreciated the collaboration with Viv Labaton, Malcolm Shephard and many others. Bob was in many ways the opposite of John Wright; intuitive rather than methodical. They sometimes came to the same conclusion from different angles and I was privy to several arguments about who had the idea first! The Culcheth group pioneered much of the experimental work and developed the understanding of the mechanism. A joint paper for the Coolant and Graphite Study Group on the inhibition by methane of the corrosion of AGL(M)P and BAEL GC(M)B was published in 1971 (Dominey *et al.*, 1971). In this paper, an equation was proposed giving the rate of methane destruction for design purposes and I was somewhat surprised that this was still being quoted when I returned to the field some 30 years later.

Harwell was a wonderfully stimulating place to work in the early 1960s. So many projects were proceeding in parallel; the fast reactor, the molten salt reactor, nuclear fusion among others; and there was always an expert you could consult on almost any scientific subject. During my employment there were four Directors, the last of whom was Lord Marshall. Walter Marshall realised that Harwell had to become more commercial and began a programme of 'diversification' which directed staff to apply their talent and experience to working with British industry. This diversification resulted in some odd assignments. My colleague, Ron Anderson, undertook the task of trying to soak up the oil which leaked from the Torry Canyon when it went aground. The history of Harwell has been recorded in the recent book by Nick Hance (2006) which I recommend.

In 1966, I joined the CEGB research laboratory at Berkeley where a Gamma Cell Co-60 facility was in the process of commissioning by Peter Jennings' team. Peter had been the CEGB representative on the various committees set up to co-ordinate graphite research carried out by the UKAEA, CEGB and consortia and I knew him well. Unfortunately Peter was seriously ill and died not long after I arrived. His contribution to the analysis and interpretation of the wide variety of work in progress at that time was considerable. The team at Berkeley carried out a study of the reaction of C-14 labelled graphites with a variety of coolants and also collaborated with in-reactor experiments mainly at Bradwell with the enthusiastic support of the station chemist, Bill Godfrey, and Peter Phennah, the GDCD engineer responsible for applying the results to reactor design.

The operation of the Gamma Cell was not always smooth. Complicated interlocks were necessary to prevent exposure of people to radiation. These were designed to fail safe and sometimes left us in a quandary when they did operate. The Cobalt-60 rods were driven in mechanically by electric motors using thick cables which had to go round several bends and of course they sometimes got stuck with the rods half way in or out. Brian Knight managed to find ways round these problems with considerable ingenuity. Initially we had no computers to aid the analysis. By 1968 we had a Mathatron electronic calculating machine which operated with punched tape. This was used by all the researchers in Materials Division and was in great demand. This led to people arriving earlier and earlier in the morning to get first use. Many of our runs went on all night. Eventually, we bought a PDP8 for the exclusive use with the Gamma cell which we had to learn to program ourselves.

During that time at Berkeley, work was in progress on the High Temperature Reactor in collaboration with the Dragon Project at Winfrith. Tim Swan, Kay Simpson, *et al.* (1970) studied the so called amoeba effect, the migration of the UO_2 core within the various layers of the HTR fuel particle. This is still the probable basis for the fuel for current designs. At that time we were expecting the first UK HTR being built at Oldbury, but this came to nought.

Another important activity was carried out by Eric Welch and his group who characterised the physical properties of samples of graphite taken from bricks used in construction of reactors. Monitoring specimens were prepared for insertion into the reactors. A large store was constructed to retain representative bricks. Responsibility for this work eventually passed to Tony Wickham.

In 1974, I left Berkeley to join the Scientific Services Department (SSD) of the South East Region. The staff were scattered around several locations in the region and they were brought together into the new laboratory at Gravesend. The laboratory was officially opened by Herman Bondi, who was then Government Chief Scientist. Over lunch he asked me about our work and how long he should speak for. He gave a wonderful talk, without a note, and finished on the dot without looking at his watch. I still have a recording of that speech which was truly inspirational. The SSD had responsibility for assisting the Magnox stations at Bradwell, Dungeness and Sizewell and the notionally first AGR under construction at Dungeness B. I spent many hours with a succession of station managers who were all colourful characters and who taught this research boffin a thing or two about the real world of power station operation and the need for real time solutions to problems. This experience was extremely valuable during my subsequent time in the industry in different jobs.

Now nearly all the Magnox stations with which I was associated in one activity or another are being decommissioned and planned lifetimes for the AGRs are way beyond what were thought possible all those years ago. I feel very lucky to have been associated with such an exciting success story, to have worked with such wonderful people and to have observed the success of teamwork in overcoming the problems which have beset the industry. I hope I have illustrated the feel of the atmosphere in the early days and why it was all so exciting. My apologies again to those many people who contributed to the story that I haven't mentioned and for the gaps I have left unfilled but this was the story of What Little I Remember.

Acknowledgements
The author would like to thank the conference organisers for the honour to present this paper and the views expressed in this paper are those of the author alone.

References
Anderson, A. R., Best, J. V. and Dominey, D. A. (1962). Radiolysis of carbon dioxide. *J. Chem Soc.*, 3498-3503.
Anderson, A. R. and Dominey, D. A. (1968). *Radiation Research Reviews*, 1, 269.
Argonne National Laboratory News Bulletin (1962), 4, December 1962.
Dominey, D. A., Labaton, V. Y. and Phennah, P. J. (1971). CEGB Report RD/B/R1873(1971).
Dominey, D. A. & Palmer, T. F. (1963). Radiolytic exchange of carbon-14 between carbon monoxide and carbon dioxide. *Discussions Faraday Society*, 36, 35.
Dominey, D. A. and Wickham, A. J. (1971). γ-Radiation induced isotope exchange in the carbon monoxide-carbon dioxide system. Studies in silica vessels. *Trans. Faraday Soc.*, 67, 2598-2606.
Frisch, O. (1979). What Little I Remember. Cambridge University Press.
Hance, N. (2006). Harwell - The Enigma Revealed. Enhance Publishing. ISBN-10: 0955305500.
Lind, S. C. and Bardwell, D. C. (1925). *J. American Chem. Soc.*, 47, 2675.
Milligan, S. (1986). Adolph Hitler: My Part in his Downfall. Penguin ISBN-10 0140035206.
Ropp, G. A., Danby, C. J. and Dominey, D. A. (1957). *J. Amer. Chem. Soc.*, 79, 4944.
Swan, T., Simpson, K., Graham, Dominey, D. A., Crofts & Collins. CEGB Report RD/B/R/1587(1970).
Wright, J. (1981). The Coolant and Graphite Chemistry of Magnox and Advanced Gas-cooled Reactors: A Review of Studies to 1980 on the Radiolytic Oxidation of Graphite by Carbon Dioxide, UKAEA Report AERE-R 10014.

The Development and Application of a Protocol for the Validation of Core Component Condition Assessment Prediction Methods

Philip R. Maul[2], Peter R. Robinson, Paul Suckling, Mark Bradford[#], Chris Wheatley* and Iain Roberson[§]

Quintessa Limited, The Hub, 14 Station Road, Henley-on-Thames, UK
[#]Engineering Division, British Energy, Barnett Way, Barnwood, Gloucester, UK
*Technical Assurance Services, Walton House, Birchwood Park, Warrington, Cheshire, UK
[§]Frazer-Nash Consultancy Limited, Stonebridge House, Dorking Business Park, UK
Email: philipmaul@quintessa.org

Abstract

British Energy needs to be able to predict the behaviour of the core as reactors age. The primary concern is core component conditions that are directly relevant to the safety case, including graphite weight loss, bore and keyway cracking and channel deformation. However, other intermediate end points, including various graphite properties, are important as being able to predict their evolution gives confidence that the system as a whole is well understood. This paper describes the development of a formal protocol for the validation of these methods that emphasises the use of quantitative methods for comparisons between 'blind' model predictions and measurements from core inspections. A pilot application of the protocol to model predictions for graphite density and bore diameters demonstrates that this is practical for both statistically-based and physically-based models.

Keywords

Graphite properties, Model validation, Predictions

INTRODUCTION

In this paper the protocol that has been developed for the validation of CCCA (core component condition assessment) modelling methods for AGRs is described. The protocol uses experience gained in applying statistical models to brick cracking (Maul and Robinson, 2005) and is designed to be compatible with relevant international and national guidance (IAEA, 2001 and Weightman, 2000). The pilot application of the protocol is then described and the lessons learned for its future application are discussed.

AN OVERVIEW OF THE PROTOCOL

The CCCA safety case leg requires predictions of core behaviour until, as a minimum, the end of the next operating cycle, and usually beyond. These predictions are obtained with models ranging from entirely empirical (*e.g.* statistical) on the one hand to mechanistic (*i.e.* physical) on the other. The protocol is applied to the validation of these models to provide direct evidence of their predictive ability. It can be applied directly to the prediction of

---

[2] Author to whom any correspondence should be addressed.

measurable quantities, but where quantities that cannot be measured need to be considered, information for measurable quantities has to be supplemented by additional arguments.

The two key principles that apply are:

- a structured approach to model validation is used independent of the type of model(s) employed; and
- each stage of the process must be documented and capable of being audited and peer reviewed.

Here a model is taken to be any systematic procedure that transforms inputs into end points of interest. This definition is quite general, but there are a number of considerations that identify 'good' models:

- In the documentation for the model, all fundamental assumptions must be made clear and, where possible, justified. It is important that the difference between model parameters that are specified and those that can be 'tuned' or 'fitted' to the available data is made clear. The number of fitted parameters is important for evaluating model performance. Modelling assumptions may need to be revisited as part of model improvement if the model is (or becomes) incompatible with the available data.
- Models are only useful if they have predictive power. It is possible for a model to provide a good representation of existing data without necessarily being able to make useful predictions outside the range spanned by the existing data. Models for core component conditions need to be able to make predictions, and this predictive power needs to be demonstrated. Comparisons of pre-inspection predictions with the outcomes of those inspections have demonstrated the predictive power of statistical models used for bore cracking (Maul and Robinson, 2005).
- In the application of a model, variability and uncertainty (in both model inputs and outputs) need to be differentiated as far as is practicable. Uncertainties in model inputs can arise, for example, due to experimental/measurement errors (which may be systematic and/or random) and uncertainties in the specification of model parameters can arise due to limited data (generally, the more data available, the less the uncertainty in the specification of model parameters). The most important source of variability in graphite properties is generally brick-to-brick with within-brick variations being less important. The environment in which the graphite ages is also variable. Separating variability and uncertainty is often difficult (and may sometimes not be possible), but, if uncertainty and variability cannot be separated, this is likely to reduce the model's predictive power.
- Models should not just calculate 'best estimate' values of end points of interest, but provide an estimate of the range of anticipated outcomes taking uncertainty and variability into account.

There are advantages in using more than one model for any prediction in order to be able to address 'conceptual model uncertainty', which results from the possibility that more than one model may provide an adequate description of the available data. It is possible, for example, that it will not be possible to differentiate between two different descriptions of an important process. It is part of British Energy's strategy to make use of different types of model where this is practical.

The main stages of the protocol are illustrated in Figure 1 with arrows representing flows of information. Protocol activities can be split into those to be undertaken before and after new data are obtained (the 'experiment' in the figure); the activities undertaken by modellers are on the right in this figure.

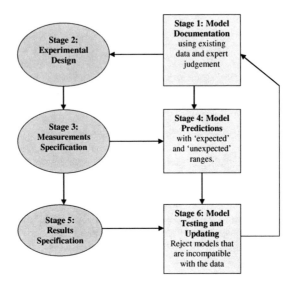

FIGURE 1: The main stages of the Validation Protocol.

Full details of the models and their parameters are documented (Stage 1) and 'blind' predictions of the experimental outcome are made (Stage 4). Model predictions provide an indication of the range of measurements that would be considered to be compatible with the model predictions and those that would be surprising, and possibly bring into question the validity of the model. The way that the predictions are presented is specific to the particular model and core component condition considered, but the general concept of differentiating between the different regions of the possible space of experimental outcomes is common. This approach has been used for several years when considering statistical models for brick cracking (Maul and Robinson, 2005), but its application to graphite property models is new.

Once the data are available, predictions and observations are compared (Stage 6); this may result in the updating of model parameters, or some models may be shown to be incompatible with the data and therefore rejected, and new models may be introduced.

An iteration of the protocol may be undertaken for a single model or for more than one type of model at the same time. Where more than one type of model is considered additional activities will be required to ensure a consistent approach to the comparison of model predictions with experimental data.

THE PILOT PROTOCOL APPLICATION

A pilot application of the protocol has been undertaken for the bore diameter and graphite density measurements obtained during the 2006 trepanning campaign at Hinkley Point B and Hunterston B. Weight loss is not measured directly, but can be related to these measured quantities. Each trepanning sample has a number of slices, generally five.

Three different types of model were applied:

- A 'trending' approach based on the use of calculations of the FEAT-DIFFUSE6 code and the concept of the Attack Rate Ratio (ARR). Models of this type are henceforth referred to as ARR models and the computer calculations are referred to simply as FEAT calculations for brevity. These models can be used to predict graphite density and weight loss. The ARR is defined as the ratio of the estimated weight loss (generally referred to as the corrected measured weight loss) to the FEAT calculation. The fundamental approach taken is to use historic data to provide corrections to the FEAT calculations that depend on Equivalent Local Burn-up (ELB), defined as the product of the core burn-up and the axial form factor at the trepanning location, as a measure of the age of the relevant brick.
- Physically-based models developed by Quintessa that also use FEAT calculations. Models of this type are henceforth referred to as Quintessa physically-based models. These models can be used to predict graphite density/weight loss and bore diameter together. In this approach the focus is on explaining the brick-to-brick variability. To do this, the bore diameter and set of graphite densities from a single trepan sample are treated as a coupled set of data, postulating a correlation between the way that the susceptibility of individual bricks to shrinkage affects the bore diameter (global shrinkage) and graphite density (local shrinkage).
- Regression models developed by Quintessa. These models can be used to predict graphite density and bore diameter separately. Regression fits for graphite density are obtained using dimensionless versions of the core burnup, the height of the sample (the vertical position in the core) and the distance of the midpoint of the sample from the inner bore edge. The forms of the regression relationships are physically motivated. It can be anticipated that regression fits to the data may become poorer if there is a change in the key processes that are operating in the core.

The models make predictions of the distribution of each of N measurements of a property of interest (denoted by z_i) that depend on the location of the sample (denoted here by r) and some measure of brick 'age' (denoted here by t). The brick-to-brick variability in graphite properties is unknown and irreducible and this limits how well any model can predict individual measurements. Because of the dependence on r and t, when a trepanning campaign is undertaken samples are not being taken from a single distribution and any measure of predictive performance needs to be based on point-by-point comparisons between predictions and measurements. A number of different methods have been employed to present the comparisons between model predictions and measurements, and two of these are illustrated here.

Outcome Categorisation Table

This scheme (illustrated in Figure 2 for one particular regression model for the Hinkley Point B density data) involves categorising where the measurements fall in the predicted distributions. In Figure 2 the following categories have been used:

+ or - indicates that the measurement was in the 50th percentile prediction interval (above or below the central model estimate);

++ or - - indicates that the measurement was in the 95th percentile prediction interval;

+++ or - - - indicates that the measurement was in the 99th percentile prediction interval; and

++++ or - - - - indicates that the measurement was outside the 99th percentile prediction interval.

The statistics of the numbers of comparisons falling into the different categories can be used to assess model performance over a number of iterations of the protocol.

Sample	Slice 1	Slice 2	Slice 3	Slice 4	Slice 5
4/L/J15	- - -	- -	-	+	+
4/L/M27	- -	-	- -	- -	- -
4/U/J15	- - -	- -	-	+	+
4/U/M27	- - - -	- -	- - - -	- - -	- - -
4/U/P19	+	-	- -	- -	- -
4/U/S35	+	+	+	+	+
5/L/J15	-	+ +	+ +	+ +	+ +
5/L/M27	+ + + +	+ + + +	+ + +	+ + +	+ +
5/L/P19	+ +	+ +	+ +	+	+
5/U/J15	- -	+	+ +	+ +	+ +
6/L/J15	+	-	+	+	+
6/U/M27	-	-	- -	- -	- -
6/U/M27	-	-	-	-	-
6/U/S35	+ + + +	+ + +	+ +	+ +	+ +
7/L/J15	- -	- -	-	-	-
7/L/M27	- - -	- -	- -	- -	- -
7/L/P19	- -	-	- -	- -	- -
7/L/S35	-	-	-	+	+
7/U/M27	- - -	- -	- - -	- - -	- -
8/L/P19	+	+	-	-	-
8/L/S35	+ +	+	+	-	-
8/U/J15	+ +	+ +	+ +	+ +	+ +
8/U/M27	+	+ +	+	+	-
8/U/P19	+	+	+	+	-
9/U/J15	+ +	+ +	+ +	+ +	+ +
9/U/M27	-	+	-	-	- -
9/U/P19	- - -	- - - -	- - - -	- - -	- - -

FIGURE 2: Outcome categorisation table for regression model HPB R1.

Validation Overview Plots

The models give a set of probability distributions (one for each measurement), and the cumulative density functions (CDFs) are denoted by F_i, $i=1,....N$. When the measurements are revealed, there is a single value, z_i, for each measurement. The position of each measurement in the predicted distribution is considered in order to obtain a set of positions, x_i, which are simply calculated as $x_i = F_i(z_i)$. Clearly the x_i lie between 0 and 1, with small values indicating cases where the measurement was at the low end of the predictions (an over-prediction) and a high value indicating that it was on the high end (an under-prediction). Plots with the derived x_i values on the x-axis and expected values on the y-axis (based on a consideration of the order statistics of size N from a uniform probability density function) have been used and referred to as Validation Overview Plots (VOPs). An example VOP is shown in Figure 3 for the Hinkley Point B density data.

A 'perfect' model would give a VOP that lies close to the line $y=x$, and it is possible to define deviations from this line that could occur by chance: the resulting confidence region is shown hatched in Figure 3.

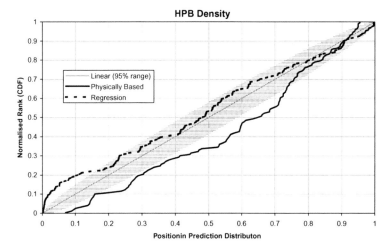

FIGURE 3: An example VOP.

Summary of Findings

The ARR and regression models performed relatively well for HPB, but not for HNB. For these models the predictions of graphite density for HPB were generally good, with the overall distribution of predictions appearing to be compatible with the distribution of measurements; the new data have improved the model fit. Figure 4 and Figure 5 show two different methods of fitting the ARR. There appears to be a trend in the evolution of ARR with ELB at HPB, but this is less apparent at HNB; this provides a possible explanation for the observed performance of these models.

By contrast, the performance of the Quintessa physically-based model was good for HNB, but poorer for HPB. This illustrates the benefits of using more than one type of model. It is too early to draw firm conclusions about the relative performances of the different types of model after only one iteration of the protocol: at least one further iteration of the protocol will be needed in order to determine whether the perceived trends are real.

The overall conclusions drawn from the pilot protocol application included:

1. No model type performed consistently better than any other: each has advantages, with all model types demonstrating useful predictive power with estimates of variability between bricks generally being good.
2. There are several advantages in taking more than one model type forward to subsequent iterations of the protocol, particularly as they will have different characteristics when extrapolated forward in time. The use of regression models is less resource intensive than other model types, and they may prove to be useful aids to judgement when assessing the performance of other, more sophisticated, types of model.

3. Valuable lessons were learned about the performance of each type of model and the need for modifications to each of these model types will be considered before the next protocol iteration.

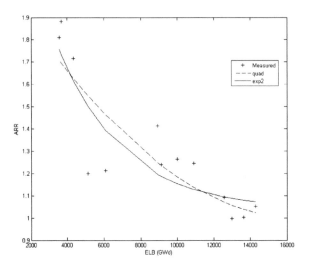

FIGURE 4: Measured and predicted ARR for HPB.

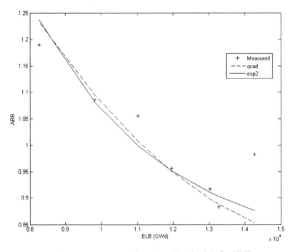

FIGURE 5: Measured and predicted ARR for HNB.

LOOKING FORWARD

The pilot protocol application demonstrated the advantages of using a formal approach to model validation and showed that it is practical to apply it to quite complex, physically based models. In particular, there is no substitute for undertaking 'blind' model predictions for reactor-based measurements in assessing the models' predictive capability. Further applications of the protocol are planned in the near future for a range of graphite properties.

In this first application of the protocol no quantitative measures were used to compare the overall predictive performance of different models, but it is planned to derive and apply such measures in future applications of the protocol. The specification of predictive performance measures is a complex issue, with most standard statistical techniques being concerned with goodness of fit measures for existing (historic) data rather than measures of predictive performance. No single measure will be satisfactory for all applications: the best measure to use will depend critically on the use to which the models are to be put.

As experience gained in using statistical models for brick cracking has shown, as further iterations of the protocol are undertaken it should be possible to make more definitive statements about the predictive powers of the different types of model and to quantify how uncertainties change with the period of time over which predictions are made.

Acknowledgements
The authors would like to thank British Energy Generation Ltd for permission to present this paper.

References
Davison, A. C. (2003). Statistical Models. Cambridge University Press.
IAEA (2001). Safety assessment and verification for nuclear power plants. Safety Guide NS-G-1.2.
Maul, P. R. and Robinson, P. C. (2007). A strategy for monitoring bore cracking in AGR graphite cores. In: *Management of Ageing Processes in Graphite Reactor Cores (Ed. G. B. Neighbour)*, Proceedings of the Ageing Management of Graphite Reactor Cores held at the University of Cardiff on 28-30 November 2005, 248-255, RSC Publishing.
Weightman, M. (2000). Nuclear Safety Directorate Technical Assessment Guide: Validation of computer codes and calculational methods.

Part B

Assessing Core Behaviour and Methodologies

Application of Whole Core Modelling Methodology with Representative Core Loading to Life Extension of AGR Graphite Cores

David J. Shaw[a], Neil McLachlan[b] and Calum Powrie[b]

[a]AMEC, Booths Park, Chelford Road, Knutsford, Cheshire, UK, WA16 8QZ.
[b]British Energy (Generation) Ltd, Barnett Way, Barnwood, Gloucester, GL4 3RS.

Abstract
The ageing of an AGR graphite core can cause distortion of the individual components, and hence geometrical configuration of the core. In addition, the ageing process of the graphite core may lead to the inducement of cracking of the fuel bricks. For life extension of the AGR reactors, it is necessary to have a whole core modelling methodology that can be used to predict the consequences of the ageing effects on the core. Such a methodology, called AGRIGID, has been developed and the methodology applied to test rig arrangements for validation purposes, and to reactor core configurations to assess future core behaviour. The loading applied to the whole core models has traditionally involved tilting the model on its side in order to achieve peak movement of the core bricks. Although the channel shapes are simple, the displacements are bounding compared to that experienced by the core under actual core loadings. It is therefore conservative to use these bounding displacements in their own right and to define complex channel shapes to assess the tolerability of the core to the predicted core distortion. In order to perform a more realistic assessment of the tolerability of the core due to ageing effects, *e.g.* increased numbers of cracked fuel bricks, a number of more representative Core Driving Loadings have been identified. The loadings have been applied to the models individually, and also cumulatively in a similar order to the time of life that they occur in the AGR core. The most significant more representative core loadings can be identified from the predicted peak core displacements, which will focus future developments. Furthermore, a comparison of these more representative core displacements can be made with the traditional bounding load core displacements in order to illustrate the contribution to the margin that exists between the bounding and the more representative approaches. This paper gives an overview of the development of the methodology, its applications and the results obtained so far.

Keywords
Finite element, AGRIGID, Representative core loading

INTRODUCTION

The ageing effects in an AGR graphite core manifest themselves in a variety of ways including physical changes corresponding to dimensional changes and ultimately distortion of the individual components and geometrical configuration of the core. In addition, the ageing process of the graphite core may lead to the inducement of through thickness full length axial cracking of the fuel bricks at keyways. Such cracking can affect the functionality of the core keying system.

For life extension of the AGR reactors, it is necessary to have a whole core modelling methodology that can be used to predict the consequences of the ageing effects on the geometrical layout of the core. A FORTRAN code has been developed and evolved over many years for the purpose of automating the production of finite element (FE) based models of core geometry used in AGR reactor core displacement behaviour studies under 'quasi-static' normal operation and fault conditions. This code is called AGRIGID and produces ABAQUS/Standard FE models incorporating rigid beam elements to represent the brick geometry and JOINTC non-linear spring elements to represent the keying arrangements of the graphite core and brick-brick interactions. The application of this methodology to test rig arrangements and reactor core configurations was given by Shaw *et al.* (2007) and McLachlan *et al.* (2007).

Many validation studies of the AGRIGID FE based models against the measured results from various test rig arrangements involving intact and cracked brick geometries have already been carried out. It has been shown that AGRIGID based models produce satisfactory correlation against test results for a variety of arrays of simulated core bricks examining behaviour on a local and global scale, when subjected to internal loading (from sources inside the array boundary) and external loading (from sources outside the array boundary or changes to the boundary itself).

TRADITIONALLY APPLIED CORE LOADING

The loading applied to the whole core models has traditionally involved tilting the model on its side in order to achieve peak movement of the core bricks. This is referred to as MDT loading (Maximum Displacement by Tilt). Although the channel shapes arising are simple, the displacements are bounding compared to that experienced by the core under actual core loadings. It is therefore conservative to use these bounding displacements in their own right and to define complex channel shapes to assess the tolerability of the core to the predicted core distortion.

DISCUSSION OF REPRESENTATIVE CORE LOADINGS

In order to perform a more realistic assessment of the justifiable tolerability of the core due to ageing effects, *e.g.* increased numbers of cracked fuel bricks, a number of more representative core loadings have been identified. A number (not exhaustive) of these are listed as follows (note that "primary" brick cracking is synonymous with "singly axial" cracked bricks):

- Manufacturing and construction tolerances – for example, the 'non-parallelness' of the end faces of fuel bricks due to tolerances can result in tilting of the fuel bricks.
- Brick end keying, rocking and eccentric features – for example, upper neutron shield bricks lean on their guide tubes, and hence apply a reaction load to the top of the core bricks below.
- Brick deformation and channel bowing – for example, differential shrinkage across a brick diameter causes the brick to assume a curved shape and hence tilt relative to its vertically adjacent neighbours.
- Brick primary crack opening – for example, after stress-reversal the cracks in singly axially cracked bricks are expected to open and such bricks may potentially interact with the neighbouring bricks.
- Core support structure tilt (otherwise known as diagrid sag) – for example, the individual tilt of the core support plates in the support structure under the weight of the core will influence the shape of the supported columns of core bricks.
- Nuclear island tilt – for example, uneven settlement of the reactor pressure vessel foundation following core build could cause a tilt and subject the core to a global lateral force proportional to the angle of tilt.

- Gas differential pressure across brick columns – for example, due to the distribution of flow passage resistances in the core, differential pressures arise which could distort the core.
- Key/keyway closure and pinching – for example, irradiation induced shrinkage reduces the core key/keyway clearances (potentially enough to completely close the clearance) leading to a stiffer core.
- Differential thermal expansion between core, core restraint structure and core support structure – for example, during a shutdown trip or a fault condition involving a rapid thermal transient a significant thermal differential may result between the core and the core restraint/support structures which could cause core distortion.

Representation of these features in the AGRIGID core modelling is discussed in the following section.

MODELLING OF REPRESENTATIVE CORE LOADINGS

The representative core loadings summarised above fall into three main categories as follows:

- Applied displacements.
- Applied forces (body forces, pressure or concentrated loads).
- End-face non-parallelness (EFNP).

The modelling of an example in each representative core loading category is given below. Note that the deformed shape plots in Figures 1 to 3 are grossly magnified for clarity, and to different scales.

Applied Displacements

The representative core loading 'differential thermal expansion between core, core restraint structure and core support structure' is incorporated into the AGRIGID FE model by applying a prescribed displacement that causes the core to distort on a shutdown trip as shown in Figure 1.

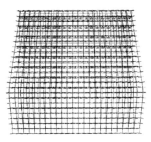

FIGURE 1: Core distortion due to differential thermal expansion of the support and restraint structure.

Applied Forces

The representative core loading 'gas differential pressure across brick columns' is incorporated into the AGRIGID FE model by applying a concentrated force that causes the core to distort as shown in Figure 2. The gas differential is greatest at the edge of the active core and causes a bowing of the channels. The peripheral bricks remain essentially vertical due to constraining effect of the core restraint.

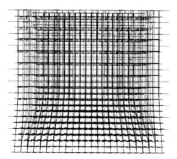

FIGURE 2: Core distortion due to gas differential pressure across brick columns.

End-Face Non-Parallelness

The representative core loading 'core support structure tilt' is incorporated into the AGRIGID FE model by applying a rotation of the boundary conditions at the base of the columns that causes the core to distort as shown in Figure 3. The dishing of the support structure is seen to cause local tilting of the lower layers of the columns.

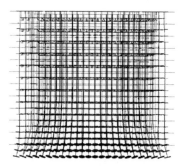

FIGURE 3: Core distortion due to core support structure tilt.

APPLICATION TO THE AGR REACTOR CORE

These representative core loadings have been analysed using AGRIGID models of the HPB/HNB core and the resulting core distortions have been quantified. The loadings are applied individually and also cumulatively in a similar order to the time of life that they occur in the AGR core. The scenarios examined for a cracked core is one with 50% of active core having singly cracked bricks.

Individual Core Loadings

A limited number of core distortion parameters have been plotted in Figure 4 in order to give an overall view of the relative importance of each loading. These are channel centre-line displacements and additional separation of horizontally adjacent bricks. The core distortion parameters have been plotted using the same scale, but this does indicate their relative importance. However, the relative importance of each of the individual representative core loadings is shown.

It can be seen that all of the loadings result in similar channel centre-line displacements and additional separation of horizontally adjacent bricks, with the exception of the scenario including differential movement of the core support and restraint structure in a fault which has larger channel centre-line displacements. The latter merely reflects the general expansion of the core in the thermal transient associated with the fault rather than a real change to the channel shape. A more relevant measure of channel shape for this scenario is the "bow" of the fuel brick centre-line, which is less than half of the centre-line displacement.

An MDT loading for the intact core has been analysed for comparison purposes and illustrates the margin that exists between the bounding and the more representative approaches when each loading is considered individually. This shows predicted core displacements considerably larger than the predicted core displacements of the representative core loadings for an intact/primary cracked core, including those of the scenario including differential movement of the core support and restraint structure in a fault when re-expressed as "bow" of the fuel brick centre-line.

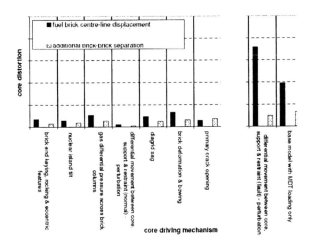

FIGURE 4: Core distortion predictions for individual core loadings.

Cumulative Core Loadings

The cumulative core distortion can similarly be plotted and the most significant representative core loadings identified. The relative importance of each of the cumulative representative core loadings can be seen in Figure 5 (though as before its does not indicate the relative importance of the core distortion parameters themselves).

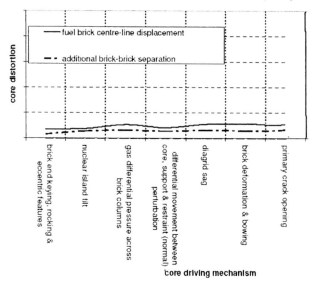

FIGURE 5: Core distortion predictions for cumulative representative core loadings.

The general level of predicted core distortion does not noticeably increase with the addition of representative core loadings, in terms of channel centre-line displacements and additional separation of horizontally adjacent bricks. The lack of notable increases in the core distortion parameters with accumulation of representative core loadings means that the distortions arising from the corresponding MDT loading, discussed under individual core loadings, remain bounding.

CONCLUSIONS

The conclusions from this study are as follows:

- The general level of predicted core distortion does not noticeably increase with the addition of representative driving loadings.

- The predicted core distortion for an intact/primary cracked core is considerably larger for the MDT loading compared to that under representative core driving loadings, either individually or cumulative.

Acknowledgements
The authors would like to thank British Energy Generation Ltd for permission to present this paper.

References
McLachlan, N., Shaw, D. J. and Salih, M. H. S. (2007). Application of whole core modelling methodology to life extension of AGR reactor graphite cores. In:- Management of Ageing in Graphite Reactor Cores (Edited by G. B. Neighbour), RSC Publishing (Cambridge), 240-247.
Shaw, D. J., Salih, M. H. S. and McLachlan, N. (2007). Development and validation of whole core modelling methodology for AGR graphite cores. In:- Management of Ageing in Graphite Reactor Cores (Edited by G. B. Neighbour), RSC Publishing (Cambridge), 256-263.

Assessment Methodologies for Simulation of Control Rod Movements in an Ageing Graphite Core

Hassan Salih[a], Calum Powrie[b] and Neil McLachlan[b]

[a]AMEC, Booths Park, Chelford Road, Knutsford, Cheshire, UK, WA16 8QZ.
[b]British Energy (Generation) Ltd, Barnett Way, Barnwood, Gloucester, GL4 3RS.

Abstract
To the meet the challenges arising from dimensional changes of the core graphite components, significant investments in both testing and model development have been made. The aim is to demonstrate satisfactory control rod (CR) functionality when operating in an aged AGR reactor core. With reactor operations, the core will undergo a degree of distortion and this may have the potential to impair the movement of the CR in the reactor core. As part of the management of the ageing effects on the reactor graphite core, an assessment methodology for simulating the interaction of a control rod with a distorted channel has been developed, validated and verified. The methodology is based on the finite element method, whereby three different modelling options have been used, namely solid-solid, beam-solid and beam-beam modelling of the CR and the channel. A full size test rig was designed and built to provide fundamental understanding and measured data for validating the modelling methodology. The paper gives an overview of the development of the methodology and its validation against results from the test rig. Cross comparison between the results from the different finite element models is also present as a verification of the different modelling approaches.

Keywords
Control rods, Ageing graphite core, Finite element modelling

INTRODUCTION

The AGR graphite components of the core undergo dimensional changes when subjected to fast neutron irradiation and such dimensional changes may impair the movements of the control rods in and out of the reactor. Hence, as part of the management of the ageing effects on the reactor graphite core, modelling methodologies have been developed to simulate the interaction of a control rod (CR) with the channel of a distorted core. The aim is to demonstrate that the geometry of the core remains functional so that the reactor can be regulated and shut down via the control rods.

Typically, an AGR core comprises layers of hollow graphite bricks, whilst a control rod is made from cylindrical tube sections coupled to each other via pin joints that allow limited articulation to take place. At the start of reactor life, the CR move in and out of the reactor core through an essentially vertical straight channel that has a bore diameter greater than the outer diameter of the CR sections.

With reactor operations, and because of irradiation, the graphite core will undergo a degree of distortion so that the initially vertically straight core channels will become distorted. The initial uniform annular gap between the outer surface of the CR and the bore of the channel will become eccentric and reduce in some locations along the length of the channel and if the channel distortion is significant, the annular gap will reduce to zero locally on one side. Under such a condition, the bore of the channel will come into contact with the outer surface of the CR and hence in addition to the weight of the CR a frictional load needs to be overcome to lift the CR. In addition, the CR will articulate and such articulation may exceed the limiting articulation of the CR joints. Under such circumstances excess stresses will arise which may impair the functionality of the CR.

Finite element based models to simulate the interaction of a CR with the bore of a distorted interstitial core channel have therefore been developed. The simulation and analysis is carried out using the general-purpose finite element code ABAQUS. This code offers many options for the finite element meshing of the geometry and the enforcement of contact of interacting structures. Three of these options have been developed, validated and verified. One such option is to represent the geometries and configurations of both the CR and the channel in detail. Such a model is a virtual computer representation and is based on the use of 3-D solid finite element modelling of both the CR and the channel.

The core functionality assessment will have to consider many different channel distorted shapes such that there is the potential for a large number of load cases to be analysed. To meet this need, two other modelling options have been developed, either of which will take less computing time than the 3-D solid modelling approach. One of these modelling approaches is based on the use beam elements to model the CR and a simpler solid elements representation for the channel. In the other, and the third modelling approach, both the CR and channel are represented by beam elements.

Whilst the use of computer simulations and in particular the use of the finite element method is widely established, there is always an added value accrued from the validation of such simulations. Towards this aim, significant investment has been made in the design and build of a test rig for the purpose of obtaining measured data to validate the FE models. The rig is full size replica of both the CR and the interstitial channel and is typical of an AGR reactor design. The data measured in this rig were the load to lift the CR and the articulations of the CR joints arising from the interaction of the CR with the bore of the channel. The CR load history and the articulations of the CR joints from the FE models were then compared with the corresponding measured data from the test rig and such comparison serves as validation evidence. The results from the different FE models have also been compared against each other to provide verification evidence of the different FE models.

CR-CHANNEL INTERACTION TEST RIG

One type of a control rod in the reactor is known as the sensor CR. A sensor CR-channel test rig, see Figure 1, was designed and operated to provide an understanding of the interaction of the sensor CR with the bore of a channel as well as to provide measured data for validating the FE models.

Test Rig Description
The test rig arrangement, Figure 1, comprises a full size replica Sensor CR and an associated channel as for the HPB/HNB reactor core. The channel comprises twelve layers of graphite bricks each of a nominal height of 825mm and a bore diameter of 127mm. The Sensor CR is a replica of a reactor design and comprises of six cylindrical sections of outer diameter of 93.34mm. The sections are of similar lengths except for the third section, which is almost three times longer. The six sections of the CR are linked to each other by articulated joints that allow a joint to rotate by approximately ±3 degrees. The overall length of the CR is ~9260mm whilst the height of the channel is ~9640mm.

A cable was attached to the upper end of the CR and a load cell was incorporated in this hoisting arrangement to enable the hoisting load to be measured during the lifting of the CR. To measure the tilting of the CR sections clinometers were internally attached to the bottom end of each CR section. The articulation of each CR joint was then deduced from the relative rotation of the two adjacent CR sections.

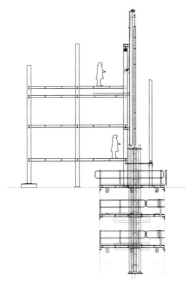

FIGURE 1: 1 CR-channel test rig.

Test Procedure
The tests in the rig involved the distortion of the channel and the subsequent insertion and removal of the Sensor CR in and out of the channel using a hoisting system that moves the CR at a slow constant speed. The distortion of the channel was achieved by laterally inclining the channel bricks in a vertical plane by jacking screws fitted near to the top and bottom ends of each brick.

Seven distorted channel shapes were simulated in the rig where the channel was essentially distorted into a number of zigzag shapes such that both the upper and lower ends of the channel were constrained in their initial positions. The channel distorted shapes used in the rig are known as shapes 14, 16, 17, 18, 19, 22 and 23; these shapes are a sub-set of 23 geometrically different shapes that were devised for the purpose investigating the effect of core distortion arising from ageing effects.

Test Rig Results
For each distorted shape, the channel was displaced at a number of shape amplitudes; the amplitude corresponding to the maximum lateral displacement of the channel shape. For each shape, the amplitude was gradually increased in magnitude in order to obtain a range of channel displaced positions and hence a range of CR-channel interactions.

For each shape and amplitude, the hoisting loads to lift the CR were measured as a function of the CR insertion movement measured from the top of the channel. Also measured are the tilting of the CR sections from which the articulations of the CR joints were obtained.

FINITE ELEMENT MODELLING AND CONTACT

The interaction of the CR with the bore of a distorted interstitial channel involves surface-to-surface contact and this is the main required modelling feature. ABAQUS (2007) is the FE code selected for the modelling and analysis of the CR contact interaction with the bore of the channel. This code has many options of simulating surface-to-surface contact with robust solution methods less prone to numerical problems.

The graphite bricks that make up the channel have been modelled using 8-node solid elements (C3D8R in ABAQUS notation). Similarly, the cylindrical sections of the CR have also been modelled using 8-node solid elements. A coarse finite element mesh was used for each of the CR and channel as there is no requirement to predict the stresses in either the CR or the channel. The contact modelling to simulate the interaction of the CR with the bore of the channel is defined via surfaces representing the outer surface of the CR and the inner surface of the channel and the pairing of corresponding surfaces. During contact, compressive forces are generated while during separation no forces are induced between the surfaces. This model is referred to as the solid-solid model, as shown in Figure 2.

A second FE model was developed in which the CR is modelled as a line model using beam elements (B31 in ABAQUS notation), whilst the channel is represented by single through thickness solid elements. For a 2-D application of this model, the channel shape is modelled as a square hollow section whilst for a 3-D model the channel will be of circular cross section and of a similar mesh to the channel in the solid-solid model. For this model, contact is enforced via the nodes of the CR beam contacting the surfaces of the channel solid model. This model is referred to as the beam-solid model, as shown in Figure 3.

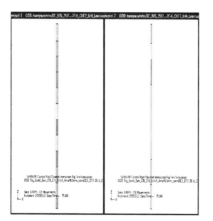   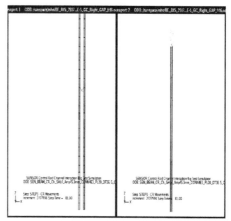

FIGURE 2: CR-channel solid-solid model.   **FIGURE 3:** CR-channel beam-solid model.

The third FE model that has been developed treats both the CR and the channel as beam models and the modelling of the interaction contact of the CR with the channel is achieved via the definition of contact surfaces associated with the nodes of the beam elements of both the CR and the channel. This model is denoted as the beam-beam model.

In all of the three FE models, the pinned joints of the CR have been represented by a combined spring-damper elements in the translational and rotational degrees of freedom. The inclusion of damping is for numerical purposes to reduce unnecessary iteration in the solution and to minimise the likelihood of numerical problems and hence increase the chances of successful solution first time.

The first two models are analysed using the 'Explicit' version of ABAQUS FE whilst the third beam-beam model is analysed using the 'Implicit' version of ABAQUS. Thus, three different models have been developed and two different solution methods used and this serves as evidence of robust verification of the three different models.

In the test rig, the lifting movement of the CR is achieved via a hoisting cable (chain), which winds up/down at a constant velocity. In the FE model this arrangement is represented by a short section of 'a simulated chain' comprising sections of truss rigid elements pinned together. The lower end of the chain model was attached to the upper end of the CR. To the upper end of the simulated chain vertical displacement is applied to simulate the vertical movement of the CR.

CR-CHANNEL INTERACTION SIMULATION

In the test rig, the CR-channel interaction is achieved by first distorting the channel into the required channel shape and then inserting/lifting of the CR in and out of the distorted channel. A similar sequence was followed in the FE simulation and analysis.

Channel Shapes Analysed

In the analysis, the channel was distorted into seven different shapes, namely shapes 14, 16, 17, 18, 19, 22 and 23, and each shape was achieved by rotation of the twelve layers of bricks of the channel relative to each other. For each channel shape, four distortions of different and increasing magnitudes have been analysed, where the peak value of the distortion is referred to as the amplitude. In the analysis, the channel was distorted in the 2-D vertical plane defined by the XZ plane where Z is the vertical axis, as the distortion of the channel in the test rig was essentially two-dimensional.

Analysis Sequence

The FE analyses of the Sensor CR and channel of the test rig were carried out in three steps:

1. In the first step, the gravity of the weight of the CR was applied.
2. In step 2, the channel distortion was applied.
3. In step 3, the lifting of the CR relative to channel was applied.

The analysis was carried using ABAQUS and terminated when the bottom end of the CR is within the top layer of the channel, after approximately eight metres of CR lifting movement.

RESULTS OVERVIEW

For each of the channel shapes analysed, the following results from the FE analysis were obtained:

- o   The undeformed and deformed shapes of the channel and the CR.
- o   Time histories of the hoisting vertical movement of the CR.
- o   Hoisting load to lift the CR.
- o   Time histories of the rotations of the top and bottom ends of CR sections.
- o   Time histories of the articulations of CR joints 1 to 4.
- o   Time histories of the predicted lateral resultant contact forces for CR sections.

The time in the histories is not real time, but a measure of the time used to simulate the hoisting movements of the CR in the analysis. Typical plots of the undeformed and deformed CR and channel shapes for a channel distorted into shape 14 and amplitude of 58.3mm are shown in Figure 4 for the solid-solid and beam-solid models.

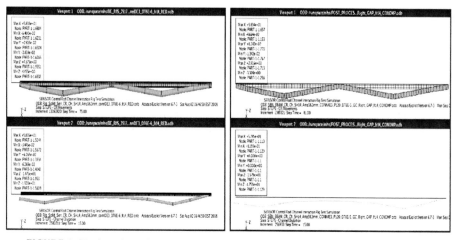

FIGURE 4: CR-channel undeformed & deformed shapes from the solid-solid and beam-solid models.

VALIDATION OF THE FE RESULTS

In the test rig, the CR hoisting load was measured using a load cell and the resulting loads for the various tests carried out were plotted as a function of the CR insertion depth as measured from the top of the channel. Also measured in the rig are the inclinations of the sections of the CR and from which the articulations of the CR joints were obtained. For the purpose of validating the FE results against the rig results, the results for particular channel shape and amplitude are compared; both visual comparison of the results in different plots and a more direct comparison of plotting both the FE and rig results in the same figure.

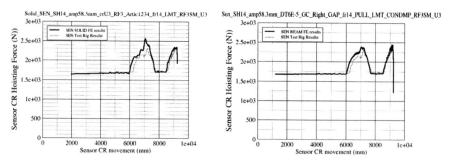

FIGURES 5 & 6: CR hoist load for shape 14 from solid-solid and beam-solid models and the test rig.

Comparison of the FE-Rig CR Hoist Loads

The results from the FE solid-solid model for the CR hoist load for shape 14 with amplitude of 58.3mm are plotted in Figure 5, whilst those from the beam-solid model are shown in Figure 6. The results from the beam-beam model for the four shape amplitudes analysed are shown in Figure 7. The corresponding results from the test rig are also shown in Figures 5, 6 and 7. From these figures, it can be seen that all of the three FE models reasonably predict profiles and peak magnitudes similar to those from the test rig. Similar correlations between the FE CR hoisting loads and the test rig values were found for the other channel shapes analysed. Hence, based on this evidence, satisfactory validation of the FE models predicted loads has been demonstrated.

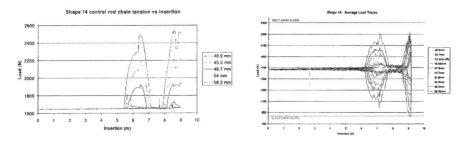

FIGURE 7: CR hoist load for shape 14 from beam-beam model (left) and the test rig.

Comparison of the FE-Rig CR Joint Articulations

In the Sensor CR-channel test rig the articulations of the CR joints 1 to 3 (joint 1 being the lowest) were plotted against the CR insertion depth, and similar results have also been obtained from the FE model. For channel shape 14 with amplitude of 58.3mm, the articulations of joint 2 are plotted in Figure 8 for the solid-solid model and in Figure 9 for the beam-solid model. For the beam-beam model, the articulation results are plotted in Figure 10, whilst the corresponding values from the test rig are also shown in Figure 8, 9 and 10. From Figures 8 to 10 it can be deduced that the three FE models have predicted similar results in profiles and magnitudes to those of the test rig. Similar correlations between the articulation results from the FE models and those from the test rig were found for the other channel shapes analysed. Hence, based on this evidence, satisfactory validation of the FE models predicted CR joint articulations has been demonstrated.

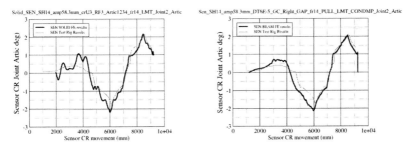

FIGURES 8 & 9: CR joint 2 articulations for shape 14 from solid-solid and beam-solid models and the test rig.

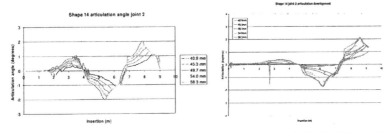

FIGURE 10: CR joint 2 articulations for shape 14 from beam-beam model (left) and the test rig.

Securing the Safe Performance of Graphite Reactor Cores

VERIFICATION OF THE FE RESULTS

Figures 5 to 7 show similarity of results for the CR hoist loads from the three (solid-solid, beam-solid and beam-beam) FE models. Furthermore, Figures 8 to 10 also show similarity in results for the CR articulations of joint 2. This similarity is evidence of satisfactory verification of the three models, which differ in FE mesh modelling, contact enforcement and solution methodology.

CONCLUSIONS

The results presented have demonstrated satisfactory validation and verification of three different methodologies for assessing the simulation of CR movements in an ageing graphite core. Hence, it can be claimed that any of the three models can serve as an analysis tool in the management of the ageing effects on AGR reactor graphite cores. The success accruing from use of these models is an essential feature of reactor lifetime extension, asset managements and safe reactor operations.

Acknowledgements
The authors would like to thank British Energy Generation Ltd and AMEC for permission to present this paper.

References
ABAQUS (2007) User's Manuals for version 6.7, Dassault Systemes Simulia Corp., Providence, RI, USA.

IMAPS – A System for Managing Graphite Core Information

Gordon J. Jahn[3] and Stephen D. J. McArthur

Dept. of Electronic and Electrical Engineering, University of Strathclyde,
204 George Street, Glasgow, UK. G1 1XW
Email: gordon.jahn@eee.strath.ac.uk

Abstract

Throughout the lifetime of the Advanced Gas-cooled Reactor (AGR) graphite cores, they have been subject to extensive inspection regimes and, more recently, monitoring has complemented the important outage inspection work to give more details about the condition of the core during reactor operation. In order to review the monitoring information, British Energy instituted Monitoring Assessment Panels and commissioned a system, the Intelligent Monitoring Assessment Panel Systems or IMAPS, to manage the information. The system is now close to deployment across the Graphite Core Project Team (GCPT) at British Energy and this paper presents the first production version of IMAPS that incorporates both monitoring and inspection information. Additionally, there is an explanation of the design process undertaken, how IMAPS contributes to AGR core information management and a demonstration of some key parts of the system. Other aspects such as data editing, auditing and validation are also considered. The paper concludes by introducing components and extensions that have been suggested and will be integrated into the system in the future.

Keywords
Condition monitoring, AGR, Multi-agent system, Intranet, World-wide web, Data dissemination

INTRODUCTION

The British AGR fleet has now been operating for over 30 years and, following life-extension for their stations, the lead-irradiation reactors are now approaching the stage where they will experience the phenomenon of stress-reversal within the cores. At this stage the neutron damage to the graphite moderator results in a reversal of compressive and tensile stresses within the graphite bricks that, along with other stresses, may cause cracks within the graphite components. The operator, British Energy, is keen to monitor the cores closely as stress reversal takes place, primarily to ensure that any core distortion, if it does happen, does not pose a threat to safety.

One difficulty in increasing the level on monitoring undertaken on AGRs is that the difficult operating conditions within the core limit the lifespan of any complex sensors. This makes it difficult to consider the addition of new on-line sensors to the reactor either in the core or within a fuel stringer and naturally suggests that the operator make the best use of any and all existing data sources in assessing the core condition.

British Energy have chosen to tackle this problem by instigating Monitoring Assessment Panels that are responsible for routine monitoring of the core data and feeding back information and recommendations to management who are able to take an informed decision about operation of those cores.

---

[3] Author to whom any correspondence should be addressed.

THE MONITORING ASSESSMENT PANELS

Monitoring Assessment Panels (MAPs) comprise a group of representatives from across the stations who meet regularly to review observations gleaned from various analytic and monitoring processes across the station and correlate the information to assess whether there is any evidence of core distortion and thereby any likelihood of there being difficulty with fuel or control rod movements. MAP meetings consider a number of Class One parameters – information from these areas are likely to be a good indicator of core distortion. These parameters are:

- Thermal-Neutron Channel Power Ratios
- Fuel Grab Load Trace (FGLT) Analysis
- Control Rod Braking Times and Control Rod Alarms
- Fuel Handling Observations

Of these, only Fuel Grab Load Trace (FGLT) is considered to be a lead indication whereby an observation might be made in advance of any safety-related core distortion. In addition to these parameters, a number of Class Two parameters are also monitored – these include environmental reports and any other station information that may be relevant.

Each of the monitoring observations is assigned a severity level which indicates how important that observation is within the context of core distortion. The severity level uses a four-point system as depicted below:

TABLE 1: MAP categorisations for monitoring observations.

Monitoring Rating Level	Description
Blue	Result indicating that there is likely to be no core distortion.
Green	Deviations from normal but that have been adequately explained and do not indicate a core safety issue.
Amber	Observations that could indicate core distortion/displacement – where explained and no core safety issue it is downgraded to green. Where unexplained remains at Amber. Where explained and a core safety issue is confirmed, upgraded to Red.
Red	Observations that unambiguously indicate core distortion/displacement

The MAP monitors the frequency of these observations looking for situations where there may be some abnormality, such as a localised cluster of amber or red observations, at which time the MAP can order a further investigation into the issue.

The remit of the MAP is to consider the current impact of all of these observations in the context of core distortion and hence safety however the challenge of handling the observations from various different sources, at varying levels across the core and potentially at varying significances made the analysis of the information difficult and complex and a solution was sought to track these items within a computerised system.

THE IMAPS SYSTEM

The Intelligent Monitoring Assessment Panel System, or IMAPS, project has been running since 2005 and was started as a research project to assist with the nascent Monitoring Assessment Panels (MAPs) at British Energy sites whilst looking to ease the task of both correlating and gathering the MAP data.

At this time, it was planned that a single system would be developed which could retrieve data from on-site systems, automatically analysing that data and informing engineers of the results. This system was to be built using a multi-agent system platform that would prove to be extensible and flexible allowing further analysis techniques and data sources to be added in the future (Jahn, 2007).

The IMAPS Prototype
Work throughout 2006 and into 2007 saw the development of a prototype system for IMAPS based fully on the agent-based approach. This system was to be the first known use of a multi-agent system for condition monitoring of a nuclear reactor and, although there were no on-line connections to additional data gathering systems, the prototype was further developed into a fully working system capable of providing rudimentary support to the MAP process with all data being entered through a web browser.

The prototype IMAPS system was provided to British Energy on a laptop and has been used at several MAPs at both Hunterston and Hinkley Point B power stations. Having this available has both aided with the MAP process and the development of IMAPS. The MAPs have been aided by quicker access to previously recorded observations whilst having a system to criticise and reflect upon has helped to guide the development of the V1.0 system.

The IMAPS prototype was demonstrated to the Health and Safety Executive's Nuclear Installations Inspectorate by British Energy at Hinkley Point B in 2007 and feedback was good, although integration with inspection information was requested. Following this, full development of a production system was started, for delivery in 2008. This would address a number of issues discovered through the prototype development and would provide the enterprise-class features required to use this system within British Energy at the appropriate data Quality-Assurance (QA) grade. The issues encountered with the prototype system are detailed in the following sections.

Applicability of Multi-Agent Systems
The original aim for IMAPS was to develop a multi-agent system to provide the necessary functionality. This approach is relatively new and relies upon a number of discrete software agents that can communicate to achieve a higher-level system goal. This approach is based upon speech act theory and by building conversations from a small number of fundamental speech acts, complex interactions between components can be built. This system should then be extensible as new agents simply need to communicate with agents that already exist within the system.

Due to the lack of condition monitoring systems built using this multi-agent system approach, a number of challenges were faced during this development. As with all multi-agent systems, the dictionary of terms and relationships used for communication – or ontology – would prove to be critical (McArthur *et al.*, 2007), but the difficulty came in the retrieval and display of the information from the system.

Whilst a key feature of multi-agent systems is that they can be very flexible in their communications, the biggest challenge encountered within IMAPS was that displaying the observations recorded on a reactor required data to be retrieved for every channel in that reactor. The prototype was designed to display a single station at a time meaning that over 800 channels (including fuel and control channels) were typically represented on-screen at any one time. In order that the interface components in the web-browser could display various different views, this meant that all data for a station was required in the browser for each user. This was neither efficient in terms of network traffic nor in terms of performance and improvements in this area were sought. This issue was compounded by the agent-based approach as, due to the underlying messaging platform between agents, it was not possible to stream data from the database to the end-user. Altogether, this proved to be a very limiting factor in the design that was a major performance bottleneck. Additionally, with software agent systems being relatively unknown, the techniques used would require extensive testing prior to use with high QA-grade data. This alone would mean that achieving a working, certified system quickly may be difficult to achieve.

Experience with the Prototype System
Whilst the multi-agent system approach resulted in some issues with the system development, there were a number of other issues to be overcome and additional requirements that needed to be met in order that the system could be used as a production system.

The first issue was that whilst capturing as much data as possible at source, analysing it and feeding all results automatically through to the MAP process would be useful; there were a number of regulatory difficulties with this proposal that would require some time to implement correctly. The most significant issue here is that the station systems operate on separate engineering networks that are not connected to the corporate network. Whilst this is a good method of preventing unauthorised access to the station network, it does make it more difficult to obtain, analyse and disseminate data.

In addition to this, the prototype relied upon its own security model with user accounts and permissions stored entirely within the application. This is useful for the prototype laptop-based installation, but could pose a real burden in rolling the system out across British Energy. At this time, it was decided that integrating IMAPS with the corporate Active Directory for authentication and permission handling would be far easier to maintain.

The final major feature that was not implemented within the prototype was a path to allow for on-line verification of all data entered. As such, one person was able to add, change and remove data without this being readily traceable. It was suggested at this point that verification be built into IMAPS V1.0 in such a way that a user cannot verify data they have entered into the system themselves.

Realising that there were a number of drawbacks and issues with the prototype, the IMAPS V1.0 development was planned and would see the key features of the prototype such as the core map, security model and the flexible data storage carried through to V1.0 for further improvement and these would be augmented with observation versioning, validation and auditing procedures so clear audit trails exist for all observations and the "current condition" on any given date can be calculated and displayed at any time.

IMAPS V1.0 DEVELOPMENT

Having developed a prototype system that was capable of supporting the MAP process, the attention turned to ruggedizing this system for use across British Energy. IMAPS V1.0 development progressed throughout late 2007 and 2008 by establishing a full set of user requirements, system design documents, architecture design and coding along with all associated testing. This development has followed the V-Model for software development which is recognised for providing robust solutions where all requirements, implementation and testing can be traced.

After a small number of MAP meetings had been undertaken at each station it was recognised that by having a reference containing all of the known conditions of the channels within the reactors, additional value can be added to the monitoring information. Whilst this was available in various company documents compiled at the time of inspection, the information was often difficult to locate quickly. Along with the feedback from the Nuclear Installations Inspectorate following the demonstration of the prototype system, this confirmed that inspection information should also be retained within the IMAPS system and from 2007, the GCPT inspection team were included in the design process allowing them to take ownership of their own information within the IMAPS system.

A key factor identified at the prototype stage was that the system should be capable of storing all the data, but also allowing for tracking validation of the data entered and checking what the suspected core condition was on any given date in the past. The state model developed and used to achieve this is described in the following section.

Observation History Storage
The four monitoring levels previously detailed in Table 1 allow for a simple yet effective prioritisation of the various observations, but it is recognised that operational decisions will need to be based upon the severity levels and observations at the time the decision is made. As such, it was recognised that a complete history for all observations should be maintained in such a way such that the "current condition" at any time in history can be viewed.

The history storage thus allows for the complete history to be maintained but ensures that the basis for any decisions can be reviewed in the future. This also ensures that it is relatively easy to trace additions, validations and changes throughout the lifetime of the system. In order to achieve this, each observation is maintained with a given state. The states are shown in Figure 1 below. The state model shown above describes the various states in which any observation can exist. It can be seen immediately that once an object is validated, that object shall never be fully removed from the system. The state model works on three main states – editable, awaiting validation and validated. New items become editable and anything in the editable state can be fully modified. When the user believes that all data has been correctly entered, they submit the observation for validation and it moves into the "awaiting validation" state. Any other user who has validation privileges can then validate this observation. The system prohibits any validation of a user's own entries. From here, the validator may either validate or return the observation for further changes. If the item is returned, it can be fully edited again whereas, if validated, the object is almost finalised within the database.

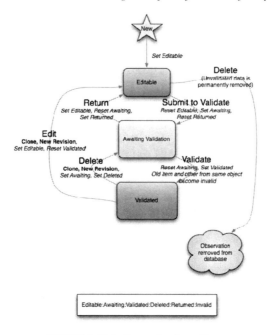

FIGURE 1: State diagram for observations.

Any changes from the validated state result in a cloning of the observation – if the object is deleted, a clone is made with a new version number and the deletion awaits validation. If the observation is to be edited, a clone is made and it moves into the editable mode whilst the existing validated observation remains "current" until the edited one is validated. Whenever an edited or deleted clone should replace the current validated version of an object, the previous version and any other items being edited that are based upon that object are flagged as invalid. Within the system, each flag is encoded as one bit of an 8-bit number with 2 spare bits. In this way, the status of every observation is represented by a number between 0 and 255.

Deployment Progress
The IMAPS system was installed in early 2009. It is deployed on a Sun Solaris 10 server and use the BE corporate Active Directory for authentication and permission management. All data will be stored within a managed Oracle database. IMAPS uses the Java Persistence API to store and retrieve the observation instances to and from the database. It is anticipated that having an easily accessible route to all graphite core information from both monitoring and inspection activity will see wider use across BE rather than solely within the GCPT. As such, the system has been designed for capacity using stateless HTTP to allow for dynamic load-balancing and multiple application servers behind front-end servers.

FUTURE WORK

The original IMAPS goal of a dynamic system performing more ongoing analysis and data-mining and pushing these results out to interested parties remains. There are a number of challenges here, not least the issue of gaining on-line access to performance data that is currently unavailable as the on-site engineering networks cannot interact with the corporate network.

Work will soon begin to look at extending this system in this way whilst further research takes place on fully validating the agent-based approach. Additional work will also commence on the intelligent processing of observations from both the "global" perspective of the consequences for each reactor and the user-specific processing that can assist and alert users when observations that may be of interest to their work are added or changed.

As well as pursuing the original means of system integration, the IMAPS system will be capable of sending information back to other systems. An expected early example of such on-line collaboration with IMAPS is the FGLT automated analysis also under development at the University of Strathclyde.

CONCLUSIONS

This paper has described the development of the British Energy IMAPS system over the last three years showing how the system has developed into the initial version available to British Energy. The paper has presented some of the key challenges experienced throughout this development and details how they have been tackled to provide a system suitable for use in the monitoring of the British Energy's reactor graphite cores.

Acknowledgements
The authors would like to thank British Energy Generation Ltd for financial support and permission to present this paper. The views expressed in this paper are those of the authors and do not necessarily represent those of the British Energy Generation Ltd. Special thanks to all those within the Graphite Core Project Team for their assistance throughout the project.

References
Jahn, G., McArthur, S. D. J. and Reed, J. (2007). An integrated architecture for graphite core data analysis. In:- *Management of Ageing Processes in Graphite Reactor Cores* (Ed. Gareth B. Neighbour), 224-231, RSC Publishing, ISBN 978-0-85404-345-3.
McArthur, S. D. J., Davidson, E. M., Catterson, V. M., Dimeas, A. L., Hatziargyriou, N. D., Ponci, F., and Funabashi, T. (2007). Multi-Agent Systems for Power Engineering Applications – Part II: Technologies, Standards and Tools for Building Multi-agent Systems, *IEEE Transactions on Power Systems*, 22, [4], Nov. 2007, 1753-1759.

On Non Smooth Contact Dynamics Methodology for the Behaviour Assessment of Discontinuous Blocky Assemblies

Nenad Bićanić[4] and Tomasz Koziara

Department of Civil Engineering, University of Glasgow,
Rankine Building, Oakfield Avenue, Glasgow G12 8LT
Email: bicanic@civil.gla.ac.uk

Abstract

Basics of the Non Smooth Contact Dynamics Method (NSCD) are reviewed and presented as a potential simulation methodology for the behaviour assessment of discontinuous blocky assemblies, with and without friction. The methodology is illustrated by lock-up simulations of the GCORE box-kite experiments series with interlocking acrylic brick assemblies and results are compared with experimental findings and the related finite element analysis.

Keywords

Multi-body dynamics, Friction, Contact

COMPUTATIONAL MODELLING OF DISCONTINUOUS BLOCKY ASSEMBLIES

Computational simulations of particulate or blocky discontinuous assemblies often need to account for a discrete nature of the inherent or evolving discontinuities. In such analyses, *'the whole'* is treated as an *'assembly of its parts'*, comprising multiple bodies, blocks or particles of arbitrary shapes (rigid or deformable), remaining, losing or potentially coming into contact as the solution progresses. There are a number of discontinuous modelling formulations, *e.g.* Discrete Element Method DEM, Modified Discrete Element Method MDEM, Combined DEM/FEM Discontinuous Deformation Analysis DDA and Non Smooth Contact Dynamics NSCD. Primary differences are reflected in the manner they deal with the characterisation of bodies geometry and deformability, the contact detection and the imposition of contact constraints and boundary conditions, as well as the associated time stepping algorithms, explicit or implicit. Further modelling features may include the definition and description of bodies fracturing and fragmentation, the transition from continua to discontinua.

Penalty Based Formulations

A number of discontinuous modelling frameworks utilise a penalty based imposition of contact interpenetration constraints. Such procedures fall into the category of *regularised* methods, as they avoid the non smooth nature of the solution, associated with jumps in the contact velocities. For example, in the frictionless contact case, the time integrator can be written in a form

$$q^{t+h/2} = q^t + \frac{h}{2} u^t \qquad [1]$$

---

4 Author to whom any correspondence should be addressed.

$$u^{t+h} = u^t + M^{-1}hf\left(q^{t+h/2}, t+h/2\right) + M^{-1}h\sum_i k_i x_i\left(q^{t+h/2}\right) \quad [2]$$

$$q^{t+h} = q^{t+h/2} + \frac{h}{2}u^{t+h} \quad [3]$$

where h is the time step, q is the configuration, u is the velocity, M is the inertia operator, f is the out of balance force, k_i is the penalty spring coefficient, $x_i(q^{t+h/2})$ is the interpenetration depth at $q^{t+h/2}$, and H_i^T transform contact forces into their generalised counterparts. The above scheme is simple to implement. Its variants, usually combined with rigid kinematics, form a basis to many discrete element codes. A shortcoming of this approach is in the necessity of using high values of penalty in order to prevent non-physical penetration. This decreases a stable time step and increases the computational cost. Moreover, in the case of rigid kinematics, the penalty coefficient becomes a "material parameter", or not quite so, as it depends on the geometry of bodies in the vicinity of a contact point.

Basics of Non Smooth Contact Dynamics (NSCD)

The NSCD method comprises distinct differences with respect to the more traditional discontinuous frameworks, as the contact conditions (unilateral Signorini condition and the dry Coulomb friction condition) are accounted for *without resorting to the regularisation* of the non smooth contact conditions. Without regularisation, due to the infinitely steep 'normal force-gap' graph associated with the Signorini condition, the solution for the actual normal contact force in a frictionless case is defined as the minimal normal force required to avoid penetration (volume exclusion requirement). Similar arguments can be developed for the frictional case as well. In the context of multiple bodies contact there are discontinuities in the velocities field and the NSCD framework recasts the entire problem in terms of *differential measures* or *distributions* with the 'non smooth' time discretised form of dynamic equations, where velocities are the primary unknowns. Considering once again the frictionless case, the Non Smooth Contact Dynamics Method (NSCD) (Moreau, 1999; Jean, 1999) formulates Equation [2] as

$$u^{t+h} = u^t + M^{-1}hf\left(q^{t+h/2}, t+h/2\right) + M^{-1}h\sum_i H_i^T R_i \quad [4]$$

where the unknown R_i is defined as the *average* contact reaction

$$R_i = \frac{1}{h}\int_t^{t+h} dR_i \quad [5]$$

The unknown local velocity (at a contact) can be further computed as

$$U_i^{t+h} = H_i u^{t+h} \quad [6]$$

Through a suitable choice of the local (contact) coordinate system it is possible to write down the following complementarity relation for the local velocity and the contact force

$$U_i^{t+h} \geq 0, R_i \geq 0, U_i^{t+h} R_i = 0 \quad [7]$$

Upon recalling that R_i and U_i^{t+h} act in the normal direction, the above condition states that for established contacts no more penetration can occur. An eventual rebounding will therefore be driven by the energy stored in a deformed body, rather then by the energy stored in a contact spring. The actual manner of energy restitution can be further elaborated by considering

$$\overline{U}_i = U_i^{t+h} + c_i min(U_i^t, 0)$$
[8]

instead of U_i^{t+h} in [7]. Here, c_i can be interpreted as a coefficient of restitution of the normal contact velocity. It can also be seen that when the local velocity $U_i^t < 0$, then $U_i^{t+h} = -c_i U_i^t$. Variables in the relation [7] can be further modified in order to account for other contact phenomena (*e.g.* cohesion). Coulomb friction can also be considered by including an additional set of implicit equations; see (Koziara, 2008) and references therein.

The NSCD equations [4, 6, 7] need to be solved for R_i and U_i^{t+h}. This in fact is the main computational expense of the method. A non-linear Gauss-Seidel scheme was originally proposed as a solution method (Moreau, 1999; Jean, 1999). Faster convergence rates can be achieved with Newton like methods (Koziara, 2008).

In summary, in the NSCD method, the contact or interface laws are treated directly, in a *non-regularised* manner. Any compliance effects are only due to the deformability of bodies. The time step is limited by the stiffest of bodies involved in a simulation. While rigid bodies can be well used in the NSCD (an example is presented in the following section), it is clear that the wave propagation effects cannot be captured without deformability. NSCD with rigid kinematics can then be used for multi-body simulations, where wave propagation effects are not essential. On the other hand, the NSCD with deformable kinematics represents a complementary framework with respect to the penalty based discrete element approaches. In the latter family of methods, contact interfaces deform, while the bodies remain rigid. In the NSCD, contact interfaces remain "rigid", while bodies deform. NSCD is usually more computationally extensive then the classical discrete element approaches. Nonetheless, it does offer a greater realism while reducing the number of necessary material parameters.

SIMULATION OF 'BOX-KITE' EXPERIMENTS

For the purpose of the so called 'box-kite' experiments, acrylic bricks were assembled into a 3x3 two-layer pattern embraced by a wooden frame (Figure 1). The middle two bricks were cracked independently at various angles (Figure 2). A manual load was applied to the two top brick halves and the maximal lock-up displacements were reported. A model of the box-kite prepared with *Solfec* software (Koziara, 2008) was used to cross-examine FE simulations as reported by Atkins (TR 5014549/06/034 Rev 0) in relation to the experimental results. Both the frictionless case and the consideration of friction were considered. The mechanical model comprised:

- Rigid kinematics.
- Ideally plastic impacts (in order to approximate quasi-static conditions of the experiment).
- Boundary conditions induced solely by contact (no explicit restrictions on displacements and rotations).
- Shear and separation loads applied directly to the mass centres of the two top brick halves (no load-induced rotation).

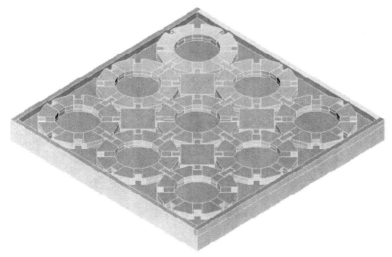

FIGURE 1: The box-kite assembly of bricks, as modelled in *Solfec*.

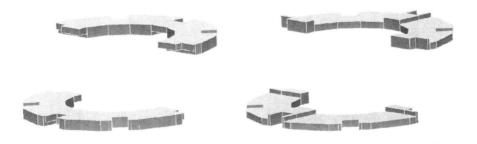

FIGURE 2: Example of cracked middle bricks from the top and bottom layers.

There were several difficulties in unambiguously validating and interpreting experimental results:

1. The force was manually applied during the experiment, an exact manner of which was not recorded.
2. The precise way by which the shearing and the separation displacements were measured during the experiment was also unknown.

The first difficulty was resolved by applying the force to the mass centres of the two top brick halves. This is equivalent to any force system whose resultant torque is zero and hence such system induces only a linear motion. Any rotations occur solely due to the contact interactions. The second difficulty has been addressed by reporting the relative displacement for a variety of selected control points. As illustrated in Figure 3, the strategy is to pick two arbitrary points A and B and allow them be convected by the motion of the respective top brick halves. The relative displacement is then measured along the fixed directions of action of the applied forces. A separate study suggested reasonable insensitivity with respect to the choice of control points, hence only one set of results, corresponding to the selection of brick halves mass centres as the control points is reported here.

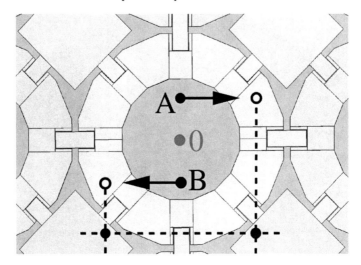

FIGURE 3: The two top brick halves and an exemplary shear displacement measurement.

The input parameters for *Solfec* simulations were

Mass density (kg / m³)	$\rho \in \{10,11\}$
Initial velocities (m/s)	*All zero*
Gravity acceleration (m/s²)	$g = [0,0,-10]$
Velocity restitution	*All zero*
Time step size (s)	$h = 0.001$
Friction coefficient	$\mu \in \{0,0.3,0.5,0.8\}$

where the smaller mass density (10 kg/m³) was used in the frictionless calculations (here irrelevant from the results standpoint, but this speeds up solution for contact reactions). When the effect of friction was investigated, the density typical for the acrylic glass was assumed (we wish those results to be easier to imagine).

Figure 4 summarises the initial set of contacts. There are no horizontal normals in the figure, because all of the bricks are initially separated by a small clearance. In the experiment, two clearance sizes were considered, referred to here as the *large* and the *small* clearance. Various orientations of crack angles correspond to different test cases, specifically numbered in the referenced reports. The labelling convention referred to the test identification number, followed by a letter, where the letters N and T correspond to the separation and shear tests. The current example should then be regarded only as a qualitative demonstration of the NSCD computational framework.

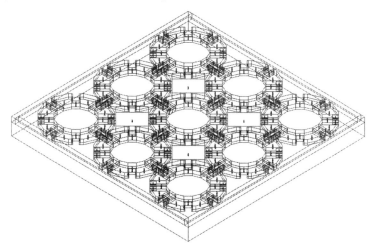

FIGURE 4: Contacts detected after the first solution time step.

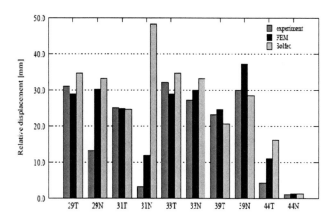

FIGURE 5: Large clearance tests - Experiment, FEM and *Solfec*.

Figures 5 and 6 compare the experimental, FEM and *Solfec* results. Both in *Solfec* and the FEM computations a zero friction was assumed. Two largest discrepancies happen for the test cases 31 and 44. Simulation of the Case 31N undergoes a complete separation, whereas the case 44T opens too wide in shear. Similarly, for the small clearance, case 48N opens too wide in separation, while case 61 opens too wide in shear. In the remaining cases NSCD results are somewhat closer to the experiment, when compared with FEM (small clearance, Figure 6).

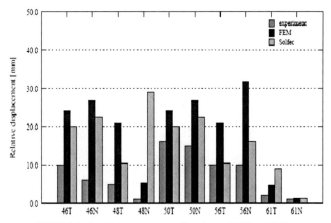

FIGURE 6: Small clearance tests - Experiment, FEM and *Solfec*.

In order to verify the role of friction, the case 31N has been given a closer look. The assumed material parameters were $\rho = 1150$ for the mass density and $\mu \in \{0, 0.3, 0.5, 0.8\}$ for friction. In the simulation, Case 31N separates fully in the frictionless case, and the purpose was to investigate whether frictional effects can affect this result (which might have happened during the experiment). The load of value 150N is ramped over the interval from 0 to 1 sec and then kept constant for another second (Figure 7). The separation is large, although the effect of friction is clear. Figure 8 shows the simulation results for an increased load of 250N, applied again over the same time interval [0,1,2]. The effect of friction is still visible, although the resulting separation is now much closer to the frictionless case.

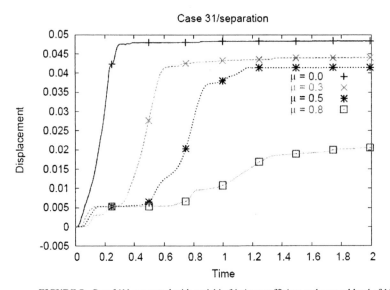

FIGURE 7: Case 31N computed with variable friction coefficient and ramped load of 150N
ramped over [0,1,2] seconds.

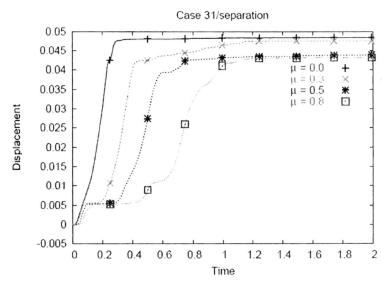

FIGURE 8: Case 31N computed with a variable friction coefficient and the load of 250N ramped over [0,1,2] seconds.

CONCLUSIONS

The Non Smooth Contact Dynamics Method (NSCD) was applied to the quasi-static analysis of lock-up scenarios for an assembly of acrylic bricks. A good overall agreement was found between the NSCD, experiments and a discrete element FEM analysis. It was also shown that friction does play a non-negligible role during the box-kite test. In Figures 7 and 8 one can notice a short plateau on the level of 5 mm, which is close to the experimental value depicted in Figure 5. It may be argued that some frictional resistance when applying the manual load during the experiment was interpreted as a lock-up.

NSCD represents a complementary tool with respect to the classical discrete element approaches. It can be applied to a variety of tasks related to the modelling of both statics and dynamics of blocky structures, with and without friction.

References
Jean, M. (1999). The non-smooth contact dynamics method. *Computer Methods in Applied Mechanics and Engineering*, 177, 235-257.
Moreau, J. J. (1999). Numerical aspects of the sweeping process. *Computer Methods in Applied Mechanics and Engineering*, 177, 329-349.
Koziara, T. (2008). Aspects of computational contact dynamics. *PhD Thesis*, University of Glasgow.

Two Dimensional Analysis of AGR Core Using a Super-Element Technique

Derek K. L. Tsang[5], Barry J. Marsden and Graham B. Heys[#]
Nuclear Graphite Research Group, School of Mechanical, Aerospace and Civil Engineering
The University of Manchester, PO Box 88, Sackville Street, Manchester, UK.
[#]Health and Safety Executive, Building 3, Redgrave Court,
Merton Road, Merseyside, UK
Email: kwong-lai.tsang@manchester.ac.uk

Abstract
A whole core model is presented that considers individual brick behaviour in addition to component interactions. The significant amount of graphite components in an AGR core makes detailed finite element analysis difficult due to the number of finite elements required and the need to take into account the contact between the many individual components. For this reason, a superelement technique has been developed to reduce the size of the analysis. In this paper, superelements for the graphite bricks are derived. The superelements are then used to model a layer of graphite bricks. The number of unknowns is reduced significantly due to smaller number of nodes. The technique has proven to be very efficient for whole core modelling.

Keywords
Whole core modelling, Super-element, Stress analysis

INTRODUCTION

A graphite core in an Advanced Gas-cooled Reactor (AGR) is a multi-layered arrangement of discrete graphite bricks that are loosely connected to each other using a system of keys located in keyways. There are two main types of bricks, of different size and geometry, namely fuel bricks and interstitial bricks. The fuel bricks are hollow in shape and are stacked in layers to form fuel channels. The interstitial bricks are also hollow and are stacked to form control rod and other utility channels. The column of bricks in each channel are keyed to the bricks in the adjacent channels using a combination of loose and integral keys located in keyways.

The ageing deformation of the graphite core structure is mainly due to irradiation induced dimensional changes and graphite material property changes causing distortion of the individual graphite components that, in turn, lead to changes to the geometrical configuration of the whole core. The irradiation induced dimensional change is a function of fast neutron fluence, irradiation temperature and radiolytic weight loss (Tsang, 2006). Therefore the component deformations are complex and depend on the position of the component in the core. The whole core distortion may also be accompanied by shrinkage induced axial cracking of some of the individual graphite fuel bricks. Furthermore such cracking may develop into double cracking, *i.e.* the fuel brick could be broken into halves. The analysis of the stresses, deformations and failure prediction of graphite moderator bricks is usually modelled by programming the irradiated graphite constitutive equations into a standard finite

---

[5] Author to whom any correspondence should be addressed.

element (FE) code using 'user material module' (Tsang, 2006). Whole core behaviour is usually modelled using specially written codes or adapted commercial codes.

Most whole core models do not explicitly model individual brick behaviour within the core. In this paper a whole core model is developed that considers individual brick behaviour in addition to the component interactions. The considerable amount of core components in an AGR makes detailed finite element analysis difficult due to the number of finite elements required and the need to take into account the contact between the many individual components. For this reason a superelement technique has been developed to reduce the size of the analysis. Using this technique the number of unknowns is reduced significantly due to the smaller number of nodal points required thus increasing the computational efficiency.

SUPERELEMENT METHODOLOGY

If repeated geometric similar parts appear in a large structure then it is appropriate to consider the use of the superelement technique. There are several advantages in using the superelement technique compared with conventional finite element method. Firstly, a superelement can be reused for representing identical structural parts such that stiffness matrix only needs to calculate once for all the identical parts. Secondly, significant effort can be saved in the procedure of finite element data input. Thirdly computational efficiency can be improved due to a large reduction in the degrees of freedom.

The procedure for applying the superelement technique consists of three steps (Tkachev, 2000). *First step: Construction of a superelement stiffness matrix.* The conventional finite element stiffness matrix **K** can be written as

$$\mathbf{Ku} = \mathbf{F} \tag{1}$$

where **u** and **F** are the displacement and force matrices, respectively. The internal finite element nodal displacements and the nodal forces are eliminated retaining only those at the boundary, *i.e.*:

$$\mathbf{u} = \begin{bmatrix} \mathbf{u}_e \\ \mathbf{u}_r \end{bmatrix} \tag{2}$$

Where subscripts '*e*' and '*r*' are eliminated and retained respectively. The stiffness matrix can be rewritten as:

$$\begin{bmatrix} \mathbf{K}_{ee} & \mathbf{K}_{er} \\ \mathbf{K}_{re} & \mathbf{K}_{rr} \end{bmatrix} \begin{bmatrix} \mathbf{u}_e \\ \mathbf{u}_r \end{bmatrix} = \begin{bmatrix} \mathbf{F}_e \\ \mathbf{F}_r \end{bmatrix} \tag{3}$$

Using the condensation technique the superelement stiffness matrix and force vector become:

$$\mathbf{K}^s = \mathbf{K}_{rr} - \mathbf{K}_{re} \begin{bmatrix} \mathbf{K}_{ee} \end{bmatrix}^{-1} \mathbf{K}_{er}$$
$$\mathbf{F}^s = \mathbf{F}_r - \mathbf{K}_{re} \begin{bmatrix} \mathbf{K}_{ee} \end{bmatrix}^{-1} \mathbf{F}_e \tag{4}$$

Hence the reduced stiffness matrix and the force vector for the superelement can be obtained:

$$\mathbf{K}^s \mathbf{u}_r = \mathbf{F}^s \qquad\qquad\qquad [5]$$

Second step: Solving the retained nodal displacement. Equation [5] is usually expressed in a local coordinate system. Therefore, all the element coordinates need to be transferred to the global coordinate system in order to assemble the global stiffness matrix, then by applying the kinematic and geometric boundary conditions, the retained node displacement can be determined.

Third step: Acquiring the eliminated nodal displacement. The eliminated nodal displacements can be calculated by

$$\mathbf{u}_e = \left[\mathbf{K}_{ee}\right]^{-1} \mathbf{F}_e - \left[\mathbf{K}_{ee}\right]^{-1} \mathbf{K}_{er} \mathbf{u}_r \qquad\qquad\qquad [6]$$

Once the eliminated nodal displacements have been determined, the strains and stresses can be conventionally computed.

SUPERELEMENTS

The condensation technique [4] is applied in the superelement method to reduce the number of unknowns for a structural part such that identical parts can be easily and repeatedly represented by a superelement. There are four different types of superelements used in the whole core modelling namely: one-eighth fuel brick superelement, one-eighth interstitial brick superelement, vertical keyway superelement and horizontal keyway superelement. Each superelement has been formed by eliminating internal nodes from the corresponding master mesh. All superelements subjected to irradiation and radiolytic oxidation conditions have been derived via the ABAQUS User Element Subroutine (UEL). Five different strains have been considered in each superelement. They are elastic, thermal, dimensional change, primary and secondary creep strains.

Fuel Brick Superelement

A superelement for a one-eighth fuel brick (FB) is shown in Figure 1a. The fuel brick superelement has 217 nodes. The corresponding master mesh used for the formulation of the superelement is shown in Figure 1b. The master mesh has 1954 nodes and 596 generalised plain strain element. A full fuel brick can be modelled by using eight superelements as shown in Figure 1c. The stiffness matrix for the superelement only needs to calculate once since there are analytical expressions for transformation, reflection and rotation of a superelement.

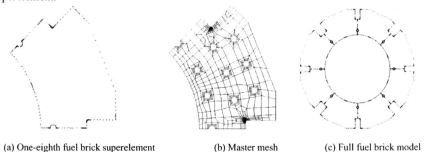

(a) One-eighth fuel brick superelement   (b) Master mesh   (c) Full fuel brick model

FIGURE 1: Fuel brick superelements.

Numerical results without a crack

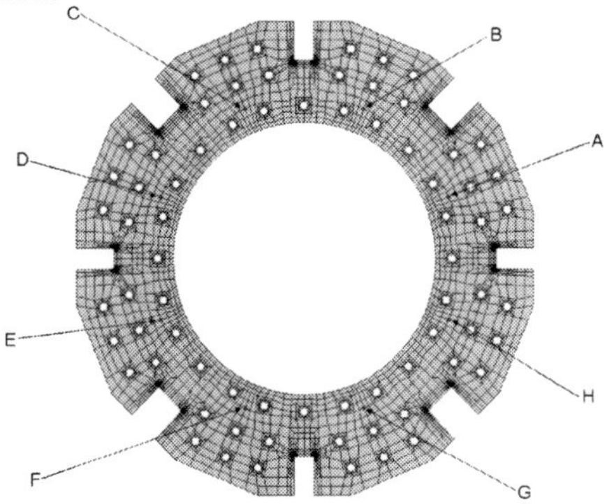

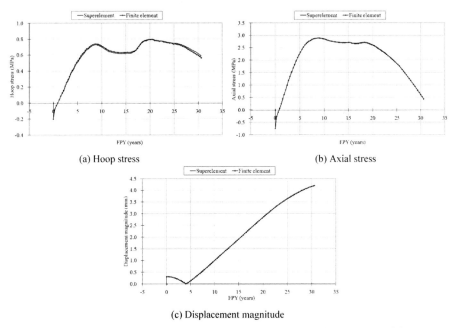

FIGURE 2: Finite element model for a full fuel brick.

A full brick has been analysed using the ABAQUS conventional finite element mesh (FE model), as shown in Figure 2, and the superelement model. Eight nodes have been selected for solution output. The locations of these nodes are shown in Figure 2.

(a) Hoop stress

(b) Axial stress

(c) Displacement magnitude

FIGURE 3: Result obtained from the superelement model and the finite element model.

Figure 3 compares the hoop stress and displacement magnitudes obtained from the superelement model with those obtained from the FE model. Figure 3c shows that the displacement magnitudes are in excellent agreement. However there are small differences in stresses. It is believed that the differences are caused by different extrapolation methods used by the ABAQUS code and the methodology described in this paper to obtain the nodal stresses from results at the integration points. The computational times for the superelement model and the FE model are 418s and 690s, respectively. The superelement model is about 40% faster than the conventional FE model.

Numerical results with a crack

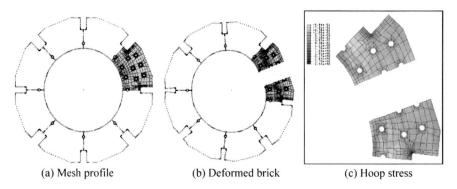

(a) Mesh profile   (b) Deformed brick   (c) Hoop stress

FIGURE 4: Superelement combined with finite elements for crack problem.

Superelements can be combined with conventional finite elements such that features like cracks can be more easily modelled. A fuel brick model combined with seven superelements and conventional finite elements is shown in Figure 4 (combined model). A pre-existing crack passing through Node A, see Figure 2, has been analysed using both the conventional FE model and the combined model. Crack faces interactions have been modelled using the ABAQUS contact analysis capability. The crack is assumed to have occurred at the start of life. At first the crack remains closed but opens later in life. The shape of the deformed brick at the end of life is shown in Figure 4b. The hoop stresses taken from the FE model at the end of life is shown in Figure 4c. The computational times for the combined model and the full finite elements model are 817.5s and 1161.9s, respectively. The combined model is about 30% faster the FE model.

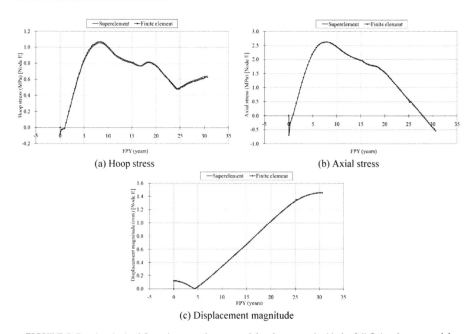

(a) Hoop stress   (b) Axial stress

(c) Displacement magnitude

FIGURE 5: Results obtained from the superelement model and compared with the full finite element model.

Figure 5a shows the comparison of the hoop stresses at Node E as in Figure 2. The results are in close agreement between the superelement model and the full finite element model. Figures 5b and 5c show the comparisons of axial stress and displacement magnitude at each node. All the results show excellent agreement between the two models.

Interstitial Brick Superelement
A superelement for a one-eighth interstitial brick and the corresponding master mesh are shown in Figures 6a and 6b. The interstitial brick superelement has 73 nodes. The corresponding master mesh has 182 nodes and 48 generalised plain strain elements. Again a full interstitial brick can be modelled by using eight interstitial brick superelements (Figure 6c). The computational times for a full interstitial brick analysis using the superelement model and the full element model are 30.7s and 65.3s, respectively. The superelement model is about 50% faster than the full element model.

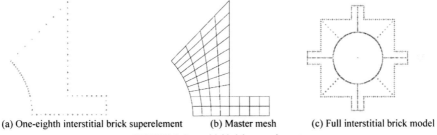

(a) One-eighth interstitial brick superelement   (b) Master mesh   (c) Full interstitial brick model

FIGURE 6: Interstitial brick superelement.

Key Brick Superelements
Superelement for horizontal key brick is shown in Figures 7a. The superelement has 113 nodes. The corresponding master mesh has 454 nodes and 132 generalised plain strain elements (see Figure 7b). The computational times for a key brick analysis using the superelement model and the full element model are 18.8s and 24.7s, respectively. The superelement model is about 25% faster than the full element model. The superelement for vertical key brick is similar to the superelement of horizontal key brick.

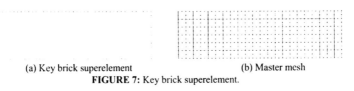

(a) Key brick superelement   (b) Master mesh
FIGURE 7: Key brick superelement.

WHOLE CORE ARRAY

In this section, four whole core arrays have been analysed using both the superelement method and the full FE model. The four whole core arrays are 2x2, 3x3, 4x4 and 5x5. The interactions between each brick components are modelled using ABAQUS gap elements. Figure 8 shows the mesh profiles for superelement model and finite element model. The element and node information for both models are given in Table 1. The number of nodes used in the superelement model is around 9 times less than the finite element model. Consequently the superelement model uses less computer memory and has faster computational times. Figure 9 shows the deformation at the end of life obtained by the superelement method and the finite element model.

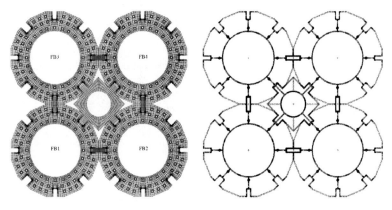

(a) Full finite element mesh   (b) Superelement mesh

FIGURE 8: Mesh profiles for core array 2x2.

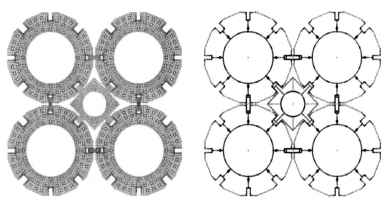

(a) Full finite element model   (b) Superelement model

FIGURE 9: Deformation at the end of life for core array 2x2.

TABLE 1: Number of elements and nodes in each array size.

Array size	Superelement model		Finite element model	
	No. of elements	No. of nodes	No. of elements	No. of nodes
2x2	144	6785	20084	64606
3x3	432	16433	46348	149077
4x4	876	30409	83564	268777
5x5	1476	48713	131732	423705

Figure 10 show the results for 2x2 array at Node A, see Figure 2, on fuel brick 1 (FB1), see Figure 8a, using the superelement model and the full finite element model. All the results show excellent agreement between the two models. Table 2 shows the computational time for different array sizes using both models. By using one CPU the superelement model is about 50% faster than the full finite element model. However, by running the models using 15 CPUs, the performance of the superelement model is faster than the finite element model for larger array sizes. Hence the superelement model is more efficient when run on a multi-processor computer than the full finite element model.

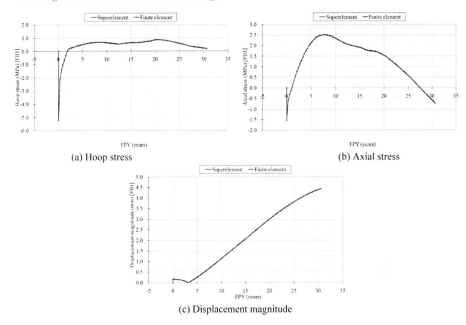

(a) Hoop stress

(b) Axial stress

(c) Displacement magnitude

FIGURE 10: Results for 2x2 array at Node A on fuel brick 1 using superelement model and full FE model.

TABLE 2: Computational times (second) by using 1 CPU and 15 CPUs for different array sizes obtained by superelement model and finite element model.

Array size	Superelement model		Finite element model	
	1 CPU	15 CPUs	1 CPU	15 CPUs
2x2	1490s	539s	3063s	740s
3x3	3335s	754s	7386s	1847s
4x4	6307s	1326s	14179s	3424s
5x5	10525s	2165s	23051s	7746s

CONCLUSIONS

Superelements have been derived for three AGR core components, a brick, an interstitial and a loose key. The analyses presented show that the superelement model and the finite element model results are in good agreement. However, the superelement model is shown to be more computationally efficient than the full finite element model as the superelement model uses less computer memory and runs faster. Also the superelement model is shown to perform more efficiently than the full finite element model on a multi-processor computer.

Acknowledgements
The authors would like to thank HSE for financial support and permission to present this paper. The views expressed in this paper are those of the authors and do not necessarily represent those of the HSE.

References
Tkachev, K.K. (2000). The use of superelement approach for the mathematical simulation of reactor structure dynamic behaviour. *Nuclear Engineering and Design*, 196, 101-104.
Tsang, D.K.L., Marsden, B.J. (2006). Development of a stress analysis code for nuclear graphite components. Journal of Nuclear Materials, 350, 208-220.

The Applicability of the Equivalent Temperature Concept for the Correlation of Graphite Dimensional Change Behaviour in Magnox Reactor Cores

Martin P. Metcalfe and John F. B. Payne

National Nuclear Laboratory, Building 102(B), Stonehouse Park, Bristol Road, Stonehouse, Gloucestershire, UK, GL10 3UT

Email: martin.p.metcalfe@nnl.co.uk

Abstract

The justification for continuing to correlate specifically PGA dimensional change data with equivalent temperature for Magnox reactor conditions is reviewed.

Keywords

Magnox, Equivalent DIDO temperature, Graphite

INTRODUCTION

Historically, "equivalent temperature" has been used to fit the temperature dependence of graphite dimensional change as a function of fast neutron irradiation. This concept is intended to allow for the effect on dimensional change and other properties of the differences in fast neutron flux between Materials Test Reactors and commercial graphite moderated power reactors. Within the UK, the concept of equivalent temperature has been adopted for the assessment of graphite property changes in materials testing reactors (MTR) and in both Magnox and AGR nuclear plant. However, the equivalent temperature "correction" that needs to be applied to true temperature becomes less significant at higher temperatures and, internationally, property changes for High Temperature Reactors (HTR) are assessed in terms of true temperature.

Recently, data for the dimensional change of AGR graphite have been successfully fitted to irradiation temperature, ignoring any effect of fast neutron flux level (Eason *et al.*, 2006). Fits to equivalent temperature for AGR graphite data have been equally successful. However, this raises the question of how successful the equivalent temperature concept is for correlating graphite property changes at Magnox operating conditions. In this paper, the justification for continuing to correlate specifically PGA dimensional change data with equivalent temperature for the lower temperature Magnox reactor conditions is reviewed.

CONCEPT OF EQUIVALENT TEMPERATURE

To arrive at the concept of "equivalent temperature", it is assumed that the mechanism by which temperature affects dimensional change is that temperature determines the probability that a collection of atoms will have the energy E needed to make a certain rearrangement of the atoms within a crystal possible. This probability, proportional to $e^{-E/kT}$ (T is the absolute temperature and k Boltzman's constant), governs the rate at which thermally activated rearrangements of collections of atoms occur at a given irradiation temperature. The rate at which atoms are displaced is proportional to the fast neutron flux, ϕ. The response of graphite to irradiation is assumed to be a function of the ratio of the rate at which atoms are displaced by irradiation, to the rate at which thermally activated rearrangements can occur at the irradiation temperature. This implies that the effect of irradiation to a given final fluence, at flux and absolute temperature ϕ_1, T_1 or flux and absolute temperature and ϕ_2, T_2, will be the same provided that:

$$\frac{\phi_1}{e^{-E/kT_1}} = \frac{\phi_2}{e^{-E/kT_2}} \text{ or, more conveniently } \frac{1}{T_2} - \frac{1}{T_1} = \frac{k}{E} \log_e\left(\frac{\phi_1}{\phi_2}\right) \qquad [1]$$

It is a convention to choose a standard position in the DIDO reactor as a reference, so if ϕ_2 is the DIDO flux, then for an irradiation at flux ϕ_1 and temperature T_1, the value of T_2 that satisfies Equation [1] is called the DIDO equivalent temperature. Alternative conventions have used standard positions in other facilities as a reference (such as Calder). As will be discussed below, different atomic rearrangements are likely to require that the atoms have different energies, E.

Calder equivalent temperature, historically adopted for the Magnox reactors, is the temperature equivalent to irradiation at a certain position in the Calder reactor, where the rating of the adjacent fuel is 3.12 MW/t$_e$ and the fast neutron flux is 3.93×10^{12} neutrons cm^{-2} s^{-1}. DIDO equivalent temperature, historically adopted for UK MTR experiments, is the temperature equivalent to irradiation at the "standard" position in the DIDO reactor, where the fast neutron flux is 4×10^{13} neutrons cm^{-2} s^{-1}. Thus, there is a one-to-one mapping between Calder equivalent temperature and DIDO equivalent temperature, which is derived from Equation [1] by putting ϕ_1 and ϕ_2 equal to the fluxes (in consistent units) at the standard positions in the Calder and DIDO reactors (Gray and Kelly, 1965):

$$\frac{1}{T_{DIDO}} - \frac{1}{T_{Calder}} = \frac{8.616 \times 10^{-5} eV\,K^{-1}}{E} \times \log_e\left(\frac{3.93 \times 10^{12}}{4 \times 10^{13}}\right) \qquad [2]$$

Since the mapping between Calder equivalent temperature and DIDO equivalent temperature is one-to-one, either temperature can be used to predict dimensional change without affecting the result, provided that one or other is used consistently throughout the analysis.

HISTORICAL EVIDENCE TO SUPPORT THE EDT CONCEPT

A number of experiments by the United Kingdom Atomic Energy Authority (UKAEA) investigated the effects of differences in neutron energy spectrum and neutron flux level between MTRs on measured property changes. Most notably, irradiation effects on dynamic Young's modulus, electrical resistivity, thermal conductivity and dimensional change were compared at the UKAEA BEPO and Danish Atomic Energy Commission DR3 facilities (Bridge et al., 1962). Effects on dimensional change and thermal conductivity were compared at the Belgian high flux MTR BR-2 and reactors of the DIDO type (Gray and Kelly, 1965). Martin and Price (1966) have also reported results from experiments on the Dounreay Fast Reactor (DFR).

Bridge et al. (1962) performed two experiments on PGA graphite. The first was designed to measure the effects on electrical resistivity of differences in neutron energy spectrum and neutron flux level between the UK BEPO reactor (maximum fast flux of less than 10^{11}

neutrons cm^{-2} s^{-1} at a normal reactor power of 6.5 MW) and the Danish DR3 reactor (maximum fast flux of 4×10^{13} neutrons cm^{-2} s^{-1} at a power output of 10 MW (the same fast flux as DIDO)). The neutron flux level experiment that followed involved a comparison of PGA graphite property changes observed on DR3 when the reactor was operated at power levels differing by a factor of ten (*i.e.* at nominal powers of 1 and 10 MW). Nominal experimental temperatures were chosen to give comparisons for the same irradiation and equivalent irradiation temperatures assuming that Equation [1] applied with E equal to 1.2 eV. Specimen irradiation temperatures were nominally 150, 182 and 219°C. Changes in Young's modulus, electrical and thermal resistivity and dimension were studied in the parallel and perpendicular directions.

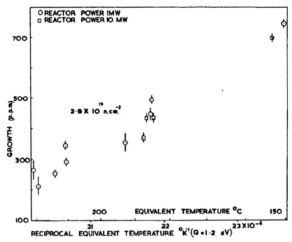

FIGURE 1: Variation of the growth with temperature for perpendicular cut PGA graphite specimens
(Bridge *et al.*, 1962).

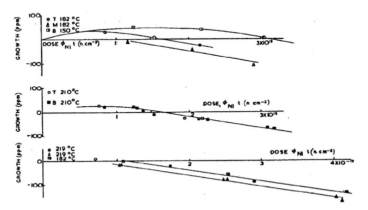

FIGURE 2: Variation of the growth with temperature for parallel cut PGA graphite specimens
(Bridge *et al*, 1962).

Figure 1 taken from the original report by Bridge *et al.* (1962) shows the dimensional change results for perpendicular cut PGA graphite specimens. By plotting the dimensional change data in terms of equivalent temperature at a reference flux of 1MW, results at the two flux levels can be trended as one population. It is clear that if all the data were plotted against irradiation temperature (for this temperature range), the observed dimensional changes for the same fluence could not be reconciled. The same effect is not apparent for parallel cut specimens owing to the weak temperature dependence over the range studied (Figure 2). Results for Young's modulus, electrical resistivity and thermal conductivity in the same DR3 experiments could also be reconciled if considered to depend on equivalent temperature, but could not be reconciled if considered to depend on irradiation temperature. The authors concluded that if the data were assessed in terms of equivalent temperature, both perpendicular and parallel results were consistent. If the data were assessed in terms of irradiation temperature, the perpendicular data could not be explained. The experiments showed that the rate of build up of damage in PGA graphite was dependent upon flux level and that this effect could be represented by the concept of equivalent temperature as described in Equation [1] with $E=1.2$ eV.

Gray and Kelly (1965) extended this earlier work to higher fluences ($\sim 10^{22}$ neutrons cm^{-2} DIDO nickel equivalent) and higher temperatures (in the irradiation temperature range 300-450°C) to provide design data for future AGRs. One experiment was run on BR-2 at 390°C and two experiments were run on the UKAEA MTRs at 350 and 450°C. Property measurements were confined to thermal conductivity and dimensional change. The dimensional change results support either choice of temperature scale. The difficulty is that the flux spectrum differs between BR-2 and DIDO, so there is uncertainty in relating the nominal fast neutron fluences of samples irradiated in the two reactors to the number of displacements per atom in the samples. Since the effect of temperature on dimensional change is small at these temperatures, it is difficult to separate effects of a different choice of temperature scale from possible differences in number of displacements per atom between samples. In contrast, there was a strong fast neutron flux level effect on thermal conductivity similar in magnitude to that expected from observations at lower temperatures.

Martin and Price (1966) performed irradiation experiments in DFR from which it is hard to draw any definite conclusions. The flux on DFR was more than 20 times greater than that of DIDO and the equivalent temperature concept may not hold under these conditions. Furthermore, no direct temperature measurements were made and uncertainties in reported values were not quantified. The temperature dependence of the DFR dimensional change data in the equivalent DIDO range 300-440°C was observed to be weak and there was little to choose between irradiation and equivalent DIDO temperature. At the lower temperatures (<300°C equivalent DIDO, <380°C irradiation), agreement between DFR and DIDO results improved when based upon equivalent temperature. The authors concluded that comparison of DFR and DIDO data showed that the DFR data were not inconsistent with equivalent temperatures. The significance of this result needs to be considered in the context of the view of Kelly (1997) that the value of the activation energy, E, would be expected to increase with irradiation temperature. Kelly also noted that attempts to determine a value of E appropriate to higher temperatures found different values for different properties or were inconclusive. Despite claims at the time and subsequent references to this series of experiments without any rigorous assessment of the actual data, MTR data provide surprisingly limited support for the trending of dimensional change behaviour in terms of equivalent temperature, although none of the data undermines the equivalent temperature concept.

ANALYSIS OF DIMENSIONAL CHANGE DATA FROM MAGNOX REACTORS

Dimensional change data from MTRs, from installed samples from ex-CEGB Magnox reactors and from measurements on Chapelcross sleeves have been analysed by Payne (unpublished in the open literature) to test whether a single expression could satisfactorily describe the observed behaviour of PGA graphite in the temperature range 250-650°C (DIDO equivalent). Whilst it is not proposed that this analysis is definitive, it succeeds in separating fluence and temperature effects present in most of the data and the resulting fits provide a convenient tool for testing the EDT concept. The method describes dimensional change as a product of temperature and fluence dependent terms:

$$\text{dimension change} = a(\theta)\left(\gamma - \gamma_{SC}\left(1 - e^{-\gamma/\gamma_{SC}}\right)\right) \qquad [3]$$

where $a(\theta)$ is the temperature term and the fluence term contains γ, the fast neutron fluence, and γ_{SC}, a fluence scale independent of temperature and orientation. The proposed expression excluded dimensional change turnaround on the basis that Magnox reactor fluences were not sufficiently high for this to occur and any onset of turnaround would be delayed by oxidation.

The form of the fluence term was based on DIDO dimensional change data obtained at nominal irradiation temperatures of 250, 300, 350, 450 and 650°C, allowing a value for γ_{SC}, the fluence scale to be determined. In addition to the evaluation of $a(\theta)$ for each DIDO temperature, γ_{SC} for both directions could be estimated. A fit to the DIDO data and a fit to the whole database showed a strong temperature dependence below 300°C and relatively little above 300°C. This is shown in Figure 3, where the data points are the values of $a(250)$, $a(300)$, $a(350)$, $a(450)$ and $a(650)$ evaluated for DIDO data and the curves represent the fits to the whole (MTR and Magnox) database. It can be seen that the choice of irradiation or equivalent temperature is important below about 300°C but above this (possibly up to ~600°C) temperature scales cannot be distinguished without more accurate data.

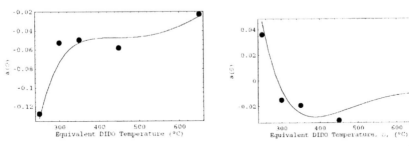

FIGURE 3: Comparison of DIDO dimensional change data parallel (left) and perpendicular (right) with temperature dependence.

As noted above, there is evidence to suggest that a higher activation energy may be more appropriate at higher temperatures and fluences, and indeed the limiting case will be where E tends to infinity and irradiation and equivalent temperature converge (see Equation [1]). Payne extended his analysis to show that DIDO equivalent temperatures calculated using an activation energy $E=3.0$ eV rather than $E=1.2$ eV made the Magnox reactor data generally less consistent with the DIDO reactor data.

RE-EVALUATION OF DIMENSIONAL CHANGE MEASUREMENTS ON CHAPELCROSS SLEEVING COMPONENTS

The dimensional change database used by Payne gave the Calder equivalent temperatures of Chapelcross sleeves, but not the irradiation temperatures. Irradiation temperatures were recently added to the database (C Mason, private communication). This allows a direct comparison of dimensional change data as a function of irradiation temperature or equivalent temperature.

The sleeve data have been grouped into two temperature ranges (230-270°C and 330-370°C) and compared with (i) DIDO data (250°C and 350°C) and (ii) fits by Payne based upon $E=1.2$ eV. Figure 4 shows perpendicular dimensional change data from DIDO (grey points) at 250°C (where irradiation temperature and equivalent DIDO temperature are the same) together with the Payne fit for the same temperature. The left hand plot shows the Chapelcross data in terms of irradiation temperature. The right hand plot shows Chapelcross data in terms of equivalent DIDO temperature. Figure 5 shows parallel dimensional change data for the same temperatures. Figures 6 and 7 repeat the analysis for the higher irradiation temperature range and with DIDO data and fits at 350°C equivalent DIDO temperature.

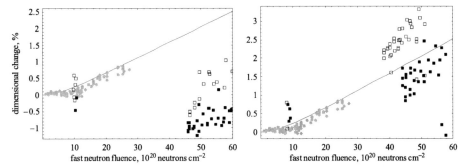

FIGURE 4: Perpendicular dimensional change data from DIDO at 250°C (grey points) together with Payne predicted fit (solid line). Chapelcross data (230-250°C open squares, 250-270°C filled squares) based upon irradiation temperatures (left hand plot) and based upon equivalent DIDO temperatures (right hand plot).

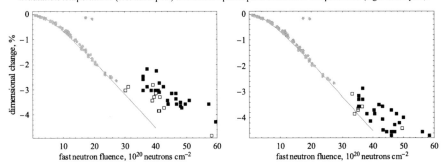

FIGURE 5: Parallel dimensional change data from DIDO at 250°C (grey points) together with Payne predicted fit (solid line). Chapelcross data (230-250°C open squares, 250-270°C filled squares) based upon irradiation temperatures (left hand plot) and based upon equivalent DIDO temperatures (right hand plot).

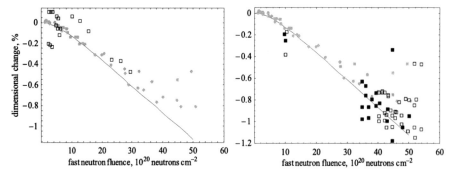

FIGURE 6: Perpendicular dimensional change data from DIDO at 350°C (grey points) together with Payne predicted fit (solid line). Chapelcross data (330-350°C open squares, 350-370°C filled squares) based upon irradiation temperatures (left hand plot) and based upon equivalent DIDO temperatures (right hand plot).

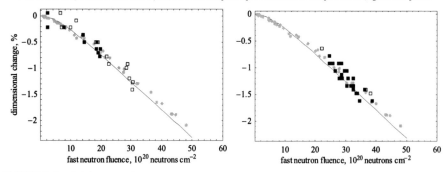

FIGURE 7: Parallel dimensional change data from DIDO at 350°C (grey points) together with Payne predicted fit (solid line). Chapelcross data (330-350°C open squares, 350-370°C filled squares) based upon irradiation temperatures (left hand plot) and based upon equivalent DIDO temperatures (right hand plot).

Chapelcross perpendicular and parallel dimensional change data for PGA graphite in the equivalent DIDO temperature range 230-270°C correlate very strongly with data from DIDO in contrast to Chapelcross data in the same irradiation temperature range (assuming an activation energy E=1.2 eV). In the equivalent DIDO temperature range 330-370°C, Chapelcross data are relatively insensitive to temperature, so a satisfactory correlation is obtained with either irradiation or equivalent DIDO temperatures.

Again, it should be noted that the Payne fits assume no dimensional change turnaround. However, the same conclusions can be drawn from Figures 4-7 on the basis of physical understanding. Since graphite crystal volume is almost unchanged by irradiation, it is clear that shrinkage of PGA graphite cannot continue indefinitely in both parallel and perpendicular directions. Considering that the crystal orientations in PGA graphite are such that the crystal swells in the perpendicular direction under irradiation, it would be expected that shrinkage in the perpendicular direction would give way to swelling at high irradiation. If the graphite is irradiated at a temperature leading to swelling in the perpendicular direction at low fluences (Figure 4), a change to shrinkage at high fluence would make no physical sense. The Chapelcross data are consistent with continuing growth perpendicular to the extrusion direction when dimensional change behaviour is correlated with equivalent DIDO temperature. In the parallel case (Figure 5), turnaround would not be expected and therefore the correlation with equivalent temperature is better explained by the data. Overall, the data would appear to be adequately represented by the Payne model.

DISCUSSION

In the irradiation temperature range 150-220°C, historical MTR data strongly support the use of equivalent temperature to explain dimensional change behaviour in PGA graphite in the perpendicular direction. Over the same temperature range in the parallel direction, no significant temperature dependence is observed, giving no evidence to discriminate between irradiation and equivalent temperature to correlate dimensional change. Similar studies carried out over the irradiation temperature range 350-450°C were inconclusive for dimensional change. Later DFR experiments carried out over the irradiation temperature range 340-805°C indicated that dimensional change observations were not inconsistent with those measured on DIDO if expressed in terms of equivalent temperature (270-340°C equivalent DIDO temperature), noting the difficulties with these experiments.

If dimensional change behaviour arising from DIDO experiments scaled according to equivalent temperature is to form the basis of Magnox reactor graphite behaviour, comparisons between DIDO and other MTRs are interesting but the acid test is the comparison with actual reactor data. The work by Payne shows that PGA graphite dimensional change data from Magnox reactors and from DIDO MTR experiments can be fitted to a single equation based upon the equivalent temperature concept (at equivalent DIDO temperatures from ~250-650°C and for fluences up to ~50×10^{20} neutrons cm^{-2}). This is based upon DIDO equivalent temperatures calculated using an activation energy E=1.2 eV, higher values of activation energy making the Magnox reactor data less consistent with the DIDO reactor data. However, the analysis shows that above 300°C equivalent DIDO temperature, the derived relationship is relatively insensitive to temperature and the data are not accurate enough to discriminate between temperature scales.

Further analysis of a sub-set of Magnox reactor data (Chapelcross) confirm that perpendicular and parallel dimensional change data in the equivalent DIDO temperature range 230-270°C (assuming an activation energy E=1.2 eV) show good agreement with data from DIDO. This contrasts with a comparison based upon irradiation temperature which shows very poor agreement. In the equivalent DIDO temperature range 330-370°C, dimensional change data are relatively insensitive to temperature.

Below equivalent DIDO temperatures of 300°C, corresponding approximately to an irradiation temperature of ~260°C in a Magnox reactor at a DIDO nickel fluence of ~50×10^{20} neutrons cm^{-2}, equivalent temperature is a better temperature scale than irradiation temperature for describing dimensional change. Dimensional change below this value is highly sensitive to small differences in temperature (Figure 3) and it is important to correlate with equivalent temperature. Above this value, dimensional change is not particularly sensitive to temperature and it is not possible to discriminate between the two temperature scales. In this case, whilst the numerical difference between irradiation and equivalent temperature will be large, the dimensional changes curves are clustered together and such large temperature differences will produce only small changes in predicted dimensional change.

This analysis supports the continued use of equivalent DIDO temperature to predict dimensional change in PGA graphite over Magnox reactor operating conditions.

Acknowledgements
The authors thank Magnox North for permission to publish. The authors wish to acknowledge their debt to Charles Mason, who re-compiled the Chapelcross sleeve data and added irradiation temperatures.

References
Bridge, H., Gray, B. S., Kelly, B. T. and Sørenson, H. (1962). An effect of flux level and flux spectrum on the accumulation of damage in reactor irradiated graphite. An Interim Report, UKAEA TRG Report 256 (C).
Eason, E. D., Hall, H. N., Marsden, B. J. and Heys, G. B. (2006). The origins and use of the equivalent temperature concept in the UK. International Nuclear Graphite Specialists Meeting, Oak Ridge National Laboratory, USA, 11th-13th September 2006.
Gray, B. S. and Kelly, B. T. (1965). A comparison of the effects of neutron irradiation on Pile Grade A graphite in BR-2 and reactors of the DIDO type. UKAEA TRG Report 886 (C).
Kelly, B. T. (1997). Irradiation damage in graphite due to fast neutrons in fission and fusion systems. Report AEAT-2391.
Martin, W. H. and Price, A. M. (1966). Determination of dose and temperature in graphite irradiation experiments in the Dounreay Fast Reactor. UKAEA TRG Report 1260 (C).

Underwriting the Weight Loss Limit for Graphite in AGR Power Stations

Peter J. Geering

Frazer-Nash Consultancy Limited, Stonebridge House, Dorking Business Park, Dorking, Surrey, RH4 1HJ, UK
Tel: +44 (0) 1306 885050
Email: p.geering@fnc.co.uk

Abstract

The UK operates a fleet of AGR (Advanced Gas-cooled Reactor) power stations. One of the plant ageing mechanisms that must be addressed is radiolytic corrosion of the graphite moderator. The action (predominantly) of gamma radiation on the carbon dioxide coolant within the pores of the graphite produces reactive species that oxidise the graphite, causing progressively increased weight losses. Currently, the limit on weight loss is defined by the extent of the experimental data available on graphite material properties. However, trepanned samples from some stations indicate that very locally the weight loss will approach this limit within the next few years. Although there is a work programme in place to use a materials test reactor to extend the database of graphite properties to higher levels of weight loss, this will not deliver data until 2010. A number of safety and operational issues have been identified which are affected by the extent of graphite weight loss. These issues relate to the structural integrity of the core, the loss of moderation, the production and consequences of debris and the mass of the core. Work is being undertaken to determine appropriate operational weight loss limits and thereby to underwrite continued safe operation of the reactors. This paper describes the work that has been undertaken to determine appropriate weight loss limits for Hinkley Point B and Hunterston B power stations. Recent work is described which has focused on the "structural integrity" issues.

Keywords
Weight loss, Limit, Safety cases

INTRODUCTION

Existing core safety cases define a limit on the weight loss of graphite, which therefore constrains the operating life of the graphite cores of AGRs. However, graphite weight loss is also used as a surrogate limit for a range of safety and operational issues and, therefore, there is the potential for a lack of clarity about the maximum acceptable level of graphite weight loss. For some stations, the current limit on graphite weight loss may be reached before the end of the station accounting life. The current limit is taken as a peak weight loss of 40% in the peak rated brick, where the weight loss is calculated using 'mean properties'. The characteristic volume over which the 'peak' is defined is taken to be that of a trepanned sample slice, *i.e.* 19mm diameter by 6mm long.

There is also a need to be confident that the stations can operate safely up to the limit of their current safety cases, and where possible extend operating lives by perhaps ten years. It is therefore necessary to have a clear understanding of what the maximum acceptable level of graphite weight loss will be for each station. The purpose of this work has been to identify a limit or limits on the level of weight loss for Hinkley Point B (HPB) and Hunterston B

(HNB) power stations within the R4 2008 Return-to-Service Safety Case and subsequent return-to-service safety cases. Some 50 safety or operational issues that may be affected by the extent of graphite weight loss have now been identified.

This paper summarises the progress that has been made in assessing the identified issues. To date assessments of eight issues, which were judged to be the most important, have been completed. The key arguments supporting the eight issues have been reviewed to see whether they currently lead to a limit on weight loss and what form this limit takes. These limits have then been compared with predicted weight losses up to the end of the period of validity of the Safety Case, to assess compliance. Progress continues to be made in assessing the remaining issues.

WEIGHT LOSS ISSUES

Some 50 weight loss issues have been identified. For clarity, the issues have been sub-divided into four categories:

- Structural integrity;
- Reactor physics and chemistry;
- Debris;
- Mass of the core.

Structural integrity is a broad category, and covers issues relating to the materials database, brick cracking, brick collapse and failure of keys and keyways. The reactor physics and chemistry issues mainly relate to the consequences of changes in moderation. The debris issues cover both mechanisms which produce debris and the consequences of the debris. Finally, there are a few issues which relate directly to the total mass of the core.

For brevity, only those issues where the assessment has been completed are discussed in this paper. The issues assessed were judged to be the most important for the following reasons. The materials property database (Issue S1) is likely to be an important constraint on the ability to assess the structural integrity of the core and hence on weight loss. Previous indications suggested that the weight loss limits relating to water ingress (Issue R3) and removal of spent fuel from the core (Issue R10) might be exceeded by the time the reactors reach the burn-up corresponding to the end of the period of validity of the safety case. Issues relating to increased neutron dose to steel components (Issue R5) and gas baffle levitation (Issue M1) are associated with important safety cases that are now in preparation. Finally, it was recognised that it was important to assess the remaining structural integrity issues to confirm the integrity of the core.

ASSESSMENT OF WEIGHT LOSS ISSUES

A summary of the assessment of eight issues is given below. A description of each issue is given, together with a brief discussion of the weight loss limit and the available evidence that supports the limit.

Issue S1: Materials Property Database
To allow graphite component integrity calculations to be performed, the variation of graphite properties with neutron dose and graphite weight loss need to be known. In particular, it is necessary to know the variation of graphite strength. Graphite material properties data have

been gathered ahead of the reactor operation in a series of Material Test Reactor (MTR) experiments with graphite of up to around 40% weight loss. The validity of applying these data to operating AGRs has been assessed and its use has been endorsed up to 40% weight loss for structural integrity assessments.

There is evidence that trends in graphite properties with weight loss do not change suddenly at weight losses above 40%. Thermally oxidised samples with weight losses well above 60% were tested in compression. Although a small number of samples were excluded from these strength tests as a result of excessive surface oxidation, in the first batch with samples ranging up to 53% weight loss only one sample was excluded. Also, it is generally understood that graphite that has been thermally oxidised suffers a greater loss of strength than graphite that has experienced radiolytic oxidation. This provides support to the argument that graphite strength does not suddenly fall at weight losses above 40%.

In addition, compressive strength data has been reported at up to 48% radiolytic weight loss for PGA (Magnox) graphite. It is generally observed that the finer the graphite grain, the stronger the material. As the Gilsocarbon graphite used in AGRs has a smaller grain size than PGA graphite, it would therefore be expected that AGR graphite will retain its strength up to at least 48% weight loss.

There is therefore evidence that there is no large discontinuity in properties (a 'cliff edge') at 40% weight loss and therefore the presence of a limited amount of material above 40% is acceptable. It is, therefore, judged that it will be tolerable that up to 2.5% of graphite at the 'brick bore' can exceed 40% weight loss. The 'brick bore' is defined in terms of a finite volume of graphite, which includes material up to a depth of 6mm from the bore surface and corresponds to the volume of a trepanned sample slice. It is important to note that this proposed limit is more conservative than the limit described in current safety cases.

Issue S4: Brick Collapse
Weight loss in the bulk of the bricks will affect their ability to support the weight of the column above. Collapse of a fuel or interstitial brick could affect its function as part of the core keyed structure and the function of the bricks above it, and hence lead to fuel and control rod channel distortions.

The stresses arising from the virgin column weight can be readily evaluated. As the fuel bricks at HPB/HNB have end face rocking features, the load at the brick interfaces will be concentrated in these regions, leading to higher localised stresses. The compressive stress at the layer 1 / layer 2 rocking features will be around 6.3MPa if evenly spread. These stresses can be compared with the strength of the graphite.

A Knudsen relationship relating strength to weight loss has been endorsed for use by British Energy. The limit of applicability of the relationship is given as 45% weight loss, although it is judged that the relationship may be applied to higher weight losses. At 55% weight loss, the Knudsen relationship gives a compressive strength of around 9MPa. This value will need to be underwritten by further work (*e.g.* MTR samples). This illustrates that there is a large margin with regard to the possibility of brick collapse as a result of evenly spread column weight. If the weight was not evenly spread, then crushing of localised areas of graphite would relieve the local stresses. Therefore a weight loss limit of 55% is applied to this issue. This limit is placed on the volumetric average weight loss of material within 6mm of the channel bore (slice 1) in the active core. The limit is conservatively placed on the slice 1 material, since the location of the load bearing region will vary depending on the deformation of the brick. The slice 1 length is based on the mean dimensions of recent trepanned specimens.

Issue S5: Rate of Oxidation Affected by Changing Gas Flow Paths due to Brick Cracking

It has been predicted that, towards the end of reactor life, the majority of fuel moderator bricks will develop single or double primary cracks, which are liable to be through wall. These primary cracks will be associated with a diameter increase, which will lead to the separation of the crack faces. Gas flow paths will then be altered, potentially leading to changes in the rate of brick oxidation. This issue investigates whether the effect of cracking in changing gas flow paths (and hence potential oxidation rates) imposes a constraint on weight loss.

Scoping calculations have been performed on Heysham 2 / Torness geometry bricks. For these calculations, a bounding case with zero pressure differential across the peak brick was considered, and the effects on weight loss at the fuel channel wall calculated. The increase in weight loss caused by cracking was predicted to be approximately 5% absolute at the fuel channel wall, and negligible at the keyway root.

The increase in weight loss predicted in the calculations above is caused by the reduction of the pressure differential between the channel bore and the arrowhead passages. Cracking of the brick will reduce this pressure differential. In the bounding case, the pressure differential will reduce from its original value to zero. It follows that the larger the change in pressure differential, the greater the change to the predicted weight loss. Heysham 2 and Torness are the only AGR stations with inter-brick sealing rings, which will result in a relatively high differential pressure. The initial pressure differential is smaller at HPB and HNB. Thus, the effect of this issue on weight loss at HPB and HNB is expected to be fairly small.

This issue only affects the rate of oxidation, which does not itself lead to a limit on weight loss. The effect on the rate of oxidation is likely to be small, and limited to the highly oxidised region at the fuel channel wall.

Issue S10: The Impact of Reactor Inspection and Trepanning Equipment on Cracked and Non-Cracked Bricks

Routine inspection of the reactors requires the lowering of items of equipment into fuel and control rod channels. This equipment includes TV inspection equipment, the Channel Bore Measurement Unit (CBMU) and trepanning equipment. These items of equipment are either required to contact, or may come into contact with, the channel wall. The area most at risk of mechanical damage from the inspection equipment will be the graphite near the fuel channel walls, where both contact and the highest weight loss occurs. The graphite at this location needs to demonstrate sufficient strength to withstand the applied loads.

It has been shown that trepanning samples from the graphite bricks is the most onerous of the inspection procedures. An assessment using a finite element model indicated that the maximum stress applied to the bricks during trepanning is approximately 2MPa. As discussed under Issue S4 above, at 55% weight loss the Knudsen relationship gives a compressive strength of around 9MPa, indicating there is a large margin on strength for this issue.

Therefore a weight loss limit of 55% is applied to this issue. This limit is placed on the volumetric average weight loss of slice 1 material in the active core. The limit is placed on the slice 1 material as this is the part of the brick supporting the inspection and trepanning loads.

Issue R3: Water Ingress

Significant boiler tube faults can have an effect on reactor moderation because water is a good moderator, and the introduction of significant quantities to the reactor, as a result of a boiler tube failure at power, has the potential to induce a reactivity fault. The effects are increased for a core which is under moderated due to graphite weight loss.

An assessment has been carried out of the effects of water ingress faults on core reactivity. This assessment assumed the graphite moderator weight loss to be ~12% of the active core. This assessment showed that the auto control system would be able to compensate for the reactivity effect of the water at the highest anticipated ingress rates by inserting the regulating rods. Therefore the limit on weight loss for this issue is currently 12% volumetric average weight loss of the active core.

Work is currently underway to assess the effects of water ingress faults into reactors with weight losses higher than have been used in previous assessments and it is expected that the limit for this issue can be increased.

Issue R5: Increased Neutron Dose to Steel Components in the Core

This issue considers the potential effects of enhanced neutron dose rates to steel components in reactor structures as a result of graphite weight loss. Core components of particular interest from a structural integrity viewpoint are the core support structure and the core restraint structure. For these components, it is desirable that the fracture toughness remains as high as possible to ensure their integrity in both normal operation and fault scenarios.

The assessment of damage dose rates at the core components does not assume a specific weight loss. Instead, weight loss has been considered as part of the uncertainty analysis. Thus, it is not yet possible to set a weight loss limit explicitly, although a limit should be capable of being defined once ongoing work to include the weight loss in the dosimetry calculations for the core restraint is complete.

Issue R10: Removal of Fuel from a Shutdown Core

During some core operations, for example to inspect the graphite bricks and other core components, a number of fuel channels may be left empty. Since highly irradiated fuel is a net absorber of neutrons, removal of such fuel may lead to a net increase in the reactivity of the core. It is necessary to ensure that Shutdown Margins are maintained with empty fuel channels in a core with graphite weight loss. The 'Shutdown Margin' is defined as the reactivity (in Niles) by which a reactor is sub-critical.

A methodology has been developed to assess the impact of empty channels on the reactivity of all station cores until the end of station life. The assessment models used assume 25% weight loss in the active core for HPB and HNB at end of station life. The Shutdown Margin is re-assessed for any fuel movements for all future outages. Therefore the limit on graphite weight loss for this issue is 25% volumetric average weight loss in the active core.

Issue M1: Gas Baffle Levitation

The gas baffle is a large cylindrical structure which separates the gas that cools the graphite from the hot gas being passed to the boilers. Up to the levitation pressure, the weight of the core (plus associated reactor internals) is sufficient to prevent the gas baffle from levitating due to the differential pressure across the boundary during both normal operation and fault conditions. Gas baffle levitation could result in an inability to trip the reactor, due to distortion of the guide tubes and possible damage to the boilers and core.

Calculations have been undertaken to predict the differential pressure for Hinkley Point B. These calculations assumed that the total mass loss of the graphite core at the end of station life was 20% of the active core and minimal, or zero, loss of mass of the neutron shield and other peripheral graphite components. The situation at Hinkley Point B is assumed also to bound Hunterston B. Since it was shown that the gas baffle will not levitate under these conditions, the weight loss limit for this issue will be taken to be 20% volumetric average weight loss in the active core.

REVIEW OF WEIGHT LOSS LIMITS

For six of the eight issues that have been assessed, a weight loss limit has been identified. This is the maximum acceptable level of graphite weight loss that the issue currently imposes on Hinkley Point B and Hunterston B power stations. These limits are summarised in Table 1 below. One issue does not lead to a limit (S5) and work is underway to identify a weight loss limit for the eighth issue (R5).

It is important to distinguish between two different types of limit. A 'direct' limit is derived directly from the issue, while an 'assessed' limit comes from an assumption that has been made during an assessment. Thus an 'assessed' limit may be relatively easy to increase by repeating the assessment assuming a higher level of weight loss, provided that acceptable results are still obtained from the new assessment. Currently, only Issue S1: Materials Property Database has a 'direct' limit.

Table 1 also investigates compliance for the R4 2008 Return-to-Service Safety Case by providing the predicted value of weight loss for both Hinkley Point B and Hunterston B power stations to the end of the period of validity proposed for the safety case. The burn-up chosen (14,235 GWd) assumes that the reactors operate at 100% power to the end of 2011. These values of weight loss have been predicted using the FEAT-DIFFUSE6 software code used by British Energy to make weight loss predictions.

The results indicate that the predicted weight losses for Hinkley Point B and Hunterston B throughout the period of validity of the R4 2008 Return-to-Service Safety Case either currently comply or should shortly comply (following further assessments) with the weight loss limits that arise from the eight issues assessed. There are two issues where further work is taking place to demonstrate full compliance. These issues are those relating to water ingress (Issue R3) and neutron dose to steel components (Issue R5).

TABLE 1: Weight Loss Limits Compliance Table
* Best Estimate

Issue Title	Weight Loss Limit		Current Prediction	
	Basis	Value	HPB	HNB
S1. Materials property database	Volumetric average weight loss of slice 1 material in the active core.	≤ 2.5% of slice 1 volume at ≥ 40%	0.06 to 0.18% > 40%	0.02 to 0.11% > 40%
S4. Brick collapse	Volumetric average weight loss of slice 1 material in the active core.	55%	0% > 55%	0% > 55%
S5. Rate of oxidation affected by changing gas flow paths due to brick cracking	No limit imposed.	-	-	-
S10. Impact of reactor inspection and trepanning equipment	Volumetric average weight loss of slice 1 material in the active core.	55%	0% > 55%	0% > 55%
R3. Water ingress	Volumetric average weight loss in the active core.	12%	13.0%*	11.7%*
R5. Increased neutron dose to steel components in the core	Likely to be based on weight losses at the core periphery.	None at present.	-	-
R10. Removal of fuel from a shutdown core	Volumetric average weight loss in the active core.	25%	13.0%*	11.7%*
M1. Gas baffle levitation	Volumetric average weight loss in the active core.	20%	13.0%*	11.7%*

CONCLUSIONS

This paper summarises the progress that has been made in identifying a limit on the level of graphite core weight loss for Hinkley Point B and Hunterston B power stations.

So far, a review of safety or operational issues that may be affected by the extent of graphite weight loss has been undertaken and some 50 issues identified. Assessments have been completed on the eight issues judged to be the most important. It is important to note that the proposed limit relating to the extent of the available database of irradiated graphite properties is significantly more conservative than the weight loss limit described in current safety cases.

The results indicate that the predicted weight losses for Hinkley Point B and Hunterston B throughout the period of validity of the R4 2008 Return-to-Service Safety Case either currently comply or should shortly comply (following further assessments) with the current weight loss limits that arise from the eight issues assessed. There are two issues where further work is taking place to demonstrate full compliance.

Work will continue to assess the remaining issues and (if applicable) to identify a weight loss limit for each issue.

Part C

Surveillance and Test Methods

BETA: A System for Automated Intelligent Analysis of Fuel Grab Load Trace Data for Graphite Core Condition Monitoring

Graeme M. West[a], Stephen D. J. McArthur[a] and Dave Towle[b]

[a]Institute for Energy and Environment, Department of Electronic and Electrical Engineering, University of Strathclyde, Royal College Building, Glasgow, G1 1XW
Email: graeme.west@strath.ac.uk

[b]British Energy Plc, Graphite Core Project Team, Design Authority, Barnwood, British Energy, Barnett Way, Barnwood, Gloucester GL4 3RS

Abstract

A key leg of the safety case required for operation of the Advanced Gas-cooled Reactor (AGR) stations is monitoring, particularly relating to the graphite core. It is not sufficient just to gather the monitoring data, but it must be analysed to extract the necessary information relating to the current condition of the core. The British Energy Trace Analysis (BETA) system has been developed to provide automated intelligent support to the analysis of Fuel Grab Load Trace (FGLT) data. FGLT data is routinely gathered during reactor core refuelling and can provide some information relating to the condition of the reactor core in addition to the inspections carried out during planned reactor outages. The process of developing the BETA software from a research prototype into a support tool which is installed on the British Energy network and accessible from anywhere in the company is described. The particular issues which have been addressed include the design, implementation, verification and validation of the software in terms of the core functionality, but also in the knowledge it contains in order to undertake its automated assessment of new refuelling event data. The second version of BETA is also described, highlighting the move to a web-based system with all the information relating to FGLT stored in a single location and describing the enhanced analysis that the BETA software undertakes. Finally, a forward look to the next version of BETA is provided, one which will integrate with other sources of condition monitoring data, and one which will have the ability to learn automatically as it is presented with new data.

Keywords
Condition monitoring, Intelligent systems, Nuclear reactor refuelling

INTRODUCTION

A key leg of the safety case required for operation of the Advanced Gas-cooled Reactor (AGR) stations is monitoring, particularly relating to the graphite core. It is not sufficient just to gather the monitoring data, but it must be analysed to extract the necessary information relating to the current condition of the core. A modern condition monitoring system permits a large volume of data to be gathered and stored, but places an increased burden on the engineer to analyse this data. Automated intelligent analysis techniques can support the analysis of condition monitoring data by providing a rapid, repeatable and auditable method of assessing the data. This allows the routine, normal, behaviour to be rapidly identified ensuring that the engineer focuses their expertise on the more unusual events. One such source of condition monitoring data is Fuel Grab Load Trace (FGLT) data. FGLT data can be interpreted to provide information relating to the condition of the graphite core bricks which comprise the reactor cores. (West, 2005) describes BETA, a software program developed to automatically assess fuel grab load trace data for the presence of anomalies. BETA has been installed on the British Energy network for over a year and is used to aid the manual analysis of FGLT data. The results of FGLT analysis are fed into regular Monitoring

Assessment Panel (MAP) meetings held at station which consider the current health of the reactor cores. As understanding of how core condition affects the FGLT increases, there has been an opportunity to improve the automated analysis. This has resulted in an improved version of BETA which is the main focus of the remainder of this paper. Firstly, though, an overview of the version 1 of BETA is provided before highlighting the differences and improvements made in the second version of the software.

BETA V1 OVERVIEW

The BETA V1 software system was developed to automatically assess FGLT data for the presence of anomalous behaviour. It was developed when electronic FGLT loggers were being introduced to Hinkley Point B and Hunterston B power stations. Anomalies in the channel wall, primarily circumferential cracks, which were known through routine CBMU (channel bore monitoring unit) inspections was shown to be appear within the FGLT traces. It was hypothesised that other types of cracks, such as those produced by both single full length axial cracks, and by double full length axial cracks could be detected, provided they resulted in a change of channel dimension. This is to say that if a single full length axial crack were to open out, to create a wider apparent channel diameter across the length of the brick or if a double cracked brick were to shear and or separate. Simple checks of the average brick layer loads were included in the first version of the BETA software, though these have been superseded by improved analysis in BETA version 2.

The main analysis performed by BETA version 1 was to automatically compare each brick layer with an expected benchmark of behaviour and report to the user any discrepancies between the two. BETA version 1 contained two main software modules, the automated analysis module and the data profiling module. The first, automated analysis, module processed the raw data, automatically identified the key features, such as brick layers, extracted the individual brick layer data and compared it to defined envelopes of expected behaviour and generated a report for the user detailing the results. Both single events and batches of refuelling events could be processed in one session by the module. The second module, the data profiling module, allowed groups of brick layer data to be viewed together and allowed benchmark behaviour to be defined from a chosen group of data. The brick layer data produced for a new, previously unseen event could be uploaded to profiling module to be included in future envelope definitions.

The development of the software was governed by a standard software development lifecycle. User requirements specification, both system and architecture design and relevant test documentation was produced to ensure that the system complied with the IT requirements. In addition, a number of tests were carried out where simulated data for hypothetical anomalies were presented to the system to demonstrate that the correct results were generated. A technical review of the software was also carried out before the software was transferred from the test environment to the live network.

BETA VERSION 2

Following installation and use in British Energy for a year, a number of suggested improvements were made to the software. Firstly, the architecture of the software was overhauled to allow the results of analysis to be retained by the system. This is an extremely important change, as it means the system becomes a single source of information relating to FGLT. This permits a number of new trending operations to be performed on the entire set of analysis results and it ensures that analyses are better managed and improves the management and maintenance of both the benchmark envelopes and the limits required by the new analysis. Secondly, the analysis performed by the system was improved. In addition to undertaking the benchmark envelope analysis described in the first system, further analysis was developed to try to identify crack configurations hypothesised by the experimental rig work.

Architecture
Version 1 of BETA was a stand alone system, where the automated module was installed on the client machine. FGLT event files were selected and processed locally, with a report containing the results of the automated analysis generated at the user's location. The segmented individual brick layer files could then be uploaded to the remote profiling module run on the server, and accessed through a web front end. Though this model was useful for proving the feasibility of the automated analysis process, there were a number of restrictions introduced by this architecture. There was no retention of the analysis results, meaning that any analysis could only draw upon results from the current session. If a batch of refuelling events were processed together, then the results say from the charge and discharge events could be compared and used to reinforce a suggested anomaly that appeared in both. However, there was no direct link to previous analyses that had been undertaken, and once the session was finished, this knowledge was effectively lost to the automated system (though was still held in the automatically generated report files). In addition, there was no way to check if the file had already been previously analysed, particularly when multiple users may be analysing FGLT events.

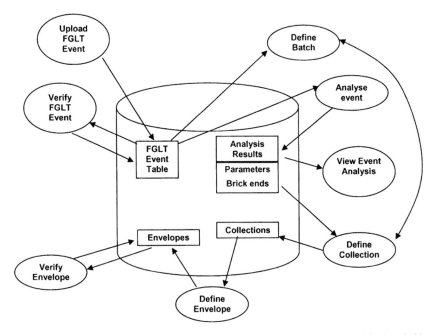

FIGURE 1: The architecture design for Version 2 of the BETA system. A central database holds all the raw FGLT data in addition to results of analysis.

The decision for version 2 of BETA was to move to a fully web-based approach, and the architecture is shown in Figure 1. A single database holds all the raw data and all the results of the analysis and acts a single repository for all information relating to FGLT data. Using this model, raw data is loaded into the database by one user and requires verification by another user before it can be analysed. Once the data is verified then it can be used in any number of analyses. This is a significant improvement over the original BETA system, where the user could select any event file to analyse, regardless of the pedigree of the data file. Though this approach does not prevent abuse, it does introduce an additional level of scrutiny of the data, requiring two independent individual to ratify a data file before it can be used. Once data has been verified for use within the system, the user can quickly perform searches and analysis on any of the data uploaded to the database. The results of any analyses are also stored within the database, so not only searches of raw data, but searches of analysis results can be performed. This means that envelopes of expected behaviour, and also associated limits required for the advanced analysis described next, can be rapidly defined.

Key Functions of BETA Version 2
As detailed in 0 there are a number of key functions performed by BETA version 2 through the web interface. The user can upload new FGLT data to the database, can verify event data from the database, can request that an event be analysed and specify the set of benchmarks used to perform the analysis. The user can view the results of any individual analysis, or view all the analyses undertaken for a selected channel. The user can search completed analyses define a collection of events. From a collection the user can select to view all the data from any one brick layer on a single graph, can view the trending of any parameter identified during the analysis and can choose to use the collection as a set of benchmark behaviour in any future analysis.

Detailed Brick Analysis
One of the largest improvements in BETA version 2 is the enhanced analysis carried out by the software. A body of work was undertaken by AMEC-NNC (Skelton, 2005) used an experimental rig to simulate the FGLT responses of various crack configurations. This work focussed on both singly and doubly axially cracked bricks at various degrees of shear and separation. Two main outcomes of this work were that the height of the brick interface peaks were affected by different crack configurations and that a step change in load across the brick layer from the expected load. Based on this work, it was determined that the shapes of anticipated defects could be determined, and therefore detected, by comparing calculations derived from five key points in each brick layer. These points are shown in Figure 2.

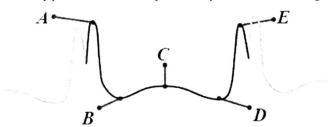

FIGURE 2: A sketch of a typical brick layer showing the 5 key brick layer points.

Using these 5 points, plus similar points from adjoining brick layers a number of base calculations can be made. The base calculations can be used to characterise features of the brick layer. For example, the difference between A and B gives the magnitude of the lower brick interface peak. The difference between point D from one brick and point B from the brick on top of it will provide an indication of a step change in load between the two layers. Data from a new trace can be assessed against expected values for each calculation to determine the presence, or otherwise, of anomalous behaviour. Table 1 shows the list of calculations performed for each brick layer.

TABLE 1: Description of the 9 base calculations.

Calculation	Description
BC1	Height of the upper brick interface peak for the previous brick layer
BC2	Difference between the upper brick interface for the previous brick layer and the lower brick interface for the current brick layer
BC3	Height of the lower brick interface peak for the current brick layer
BC4	Difference in the peak base heights between the upper brick interface for the previous brick layer and the lower brick interface for the current brick layer
BC5	Difference in the peak base heights between the upper brick interface for the current brick layer and the lower brick interface for the next brick layer
BC6	Height of the upper brick interface peak for the current brick layer
BC7	Difference between the upper brick interface for the current brick layer and the lower brick interface for the next brick layer
BC8	Height of the lower brick interface peak for the next brick layer
BC9	Size of the middle peak, C

These base calculations can be used to characterise the expected profiles of various types of anomaly. Table 2 shows some of the types of hypothetical crack configurations that can be detected. The sequence column refers to the results of the base calculations from 1 to 9. An *n* indicates normal behaviour, a + indicates abnormally large, a – indicates abnormally small, *N-* means not negative and A means any value. For example, a circumferential crack will be detected by BC1-BC8 being normal and BC9 being larger than expected.

In addition to the hypothesised anomalies shown in Table 2, if in future new anomalies are found, and the associated sequence of calculations could be determined, then this could quickly be added to any analysis undertaken. Furthermore, if it was determined that a new point was required in order to conduct the analysis, then this could be readily defined within the software and stored within the database. It would be straightforward to modify the analysis engine to determine this point for new analysis, but also very easy to configure it to search the database for completed analyses and calculate this required value for the missing events. One such example might be a circumferential crack combined with a double axial crack for the crack interface to one brick end. In this case, a suitable sequence would need to be defined which would likely be a combination of the circumferential crack and the low resistance profiles. In addition, there would probably need to be a parameter introduced which would measure the difference in mean loads for the two brick halves, say a measure of point B – Point D from Figure 2.

TABLE 2: Profiles for some possible crack configurations.

Description	Sequence	Sketch
Narrow Bore	n,+,+,+,+,n,N-,n,A	
Circumferential	n,n,n,n,n,n,n,+	
Low Resistance	n,n,+,+,+,+,n,n,-	
Possible Separation	n,n,n,n,n,n,+,+,A	
Possible Shear	n,+,+,n,n,n,n,n,A	

Defining Limits

The major issue with this approach is the same as the one faced when defining envelopes of expected behaviour, is how to set the limits of expected behaviour. In otherwise, how can it be determined if BC1 is abnormally high? This problem is further exacerbated by the natural variation in the load signal across the length of the core. Figure 2 shows an ideal brick profile, but external effects, such as addition friction introduced by the upper stabilising brushes and aerodynamic effects caused by varying gas flows through the channel mean that the base calculation limits for each layer will be required to be undertaken separately. Figure 3 illustrates this point, where there is a known step change in load in brick layer 4 and the effects of masking introduced by the upper stabilising brushes passing through a channel restriction in brick layer 5. Brick layer 6 is closest to the ideal shape shown in Figure 2, though even here, the middle peak is not so prominent.

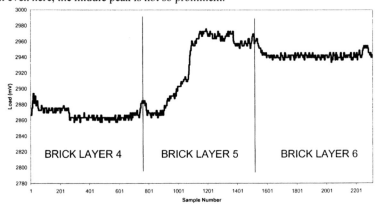

FIGURE 3: A plot showing brick layers 4, 5 and 6.

The chosen approach is to base the limits for the base calculations on previous events. The volume of limits required to be defined makes it impractical to undertake a manual assessment for each limit. Nine base calculations for each of the 10 brick layers separate for both charge and discharge means 180 upper and 180 lower limits are required. Instead a collection of refuelling event data is defined, which has been verified to contain only examples of normal behaviour. The base calculations for this collection of data are then used to statistically define the upper and lower bounds.

In the first version of BETA, the envelopes were defined with no reference to the load conditions within the reactor. It has been shown recently (Wallace, 2008) that reactor parameters can affect the apparent weight of the fuel stringer. Therefore, it would be prudent to define different limits for different reactor conditions. Again, as a better understanding of how reactor parameters affect the FGLT is gained, this can be factored into the definition of the limits for expected behaviour.

The approach for defining an upper and lower limit is consistent for all the calculations. Each base calculation is assumed to have a Gaussian distribution, and the limits set to be the standard measure for defining outliers as set out by (Tukey, 1977). That is, the upper limit is 1.5 times the inter-quartile range added to the upper quartile value and the lower limit is the lower quartile minus 1.5 times the inter-quartile range. BETA version 2 presents the user with a plot of all the raw parameter values in a collection, along with a plot of the distribution of this data to allow the user to determine whether the collection is suitable to be used as a benchmark. Figure 4 shows a sample output for base calculation 9 for a collection of 63 brick layer 6 charge events. In this particular example, it is noted that the distribution does not follow a single Gaussian, but instead appears to be composed of two sub-populations. The data presented here is intended as an illustrative example, and questions would be raised as to why there are two separate distributions noted here.

Automated Learning of Limits
Currently the definition of these limits is still undertaken manually and once they are defined they are fixed in time. In the next version of the BETA software, it is the vision that as new analysis are performed with the system, it will have an element of self-awareness which should allow it to utilise the results of the analysis to update its own understanding of normal behaviour. In other words, if a new event is analysed and was found to be OK, it could use the data from this event to update its own limits. This way the limits could be kept up to date and adapt to longer term trends as the core ages further. In order to achieve this, the system will need to be provided with knowledge of what right and wrong is, and it is envisaged that this information will be gained from the IMAPS system. The IMAPS system (Jahn, 2007) is designed to support the MAP meetings, and records information relating to all of the channels discussed during a MAP and includes inspection as well as condition monitoring data. Access to this verified source of data would allow the system to automatically decide whether FGLT events should be included in the definition of limits. There are still very difficult questions to be tackled with regards the verification and validation of such an approach, in particular how could it be guaranteed that these learned limits are indeed valid? In order to start to tackle these issues, however, it was critical that the BETA software gained the ability to store and access all information relating to FGLT.

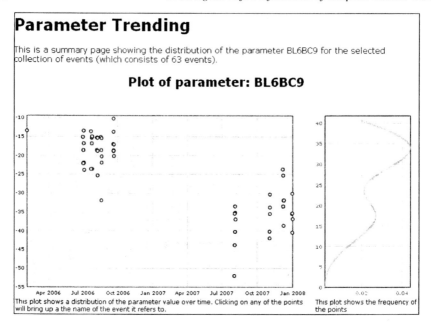

FIGURE 4: A sample screenshot from BETA version 2 showing the trending of the BC9 parameter for brick layer 6 along with the frequency distribution plot.

Deployment

Version 2 of the software is undergoing installation on the test environment at British Energy. There it will be tested and undergo a technical review before being released onto the live system.

CONCLUSIONS

This paper has provided an overview of the second version of the BETA system. It has described the improvements made over the initial version of the system, in terms of the enhanced analysis, but also in the way that it allows all the information relating to FGLT to be managed. Details of possible future enhancements to the software are also detailed, moving towards a system which is able to manage and maintain its own model of normal behaviour in order to maintain accurate assessments.

Acknowledgements
The authors would like to thank British Energy Generation Ltd for financial support and permission to present this paper. In particular thanks to Jacob Smith for his contributions to the definition of the various crack configurations. The views expressed in this paper are those of the authors and do not necessarily represent those of the British Energy Generation Ltd.

References
Jahn, G., McArthur, S. D. J. and Reed, J. (2007). An integrated architecture for graphite core data analysis. In: *Management of Ageing Processes in Graphite Reactor Cores (Ed. G. B. Neighbour)*, Proceedings of the Ageing Management of Graphite Reactor Cores held at the University of Cardiff on 28-30 November 2005, 224-231, RSC Publishing.
Skelton, J. (2005). Fuel grab load trace applied to the detection of cracks in AGR fuel bricks: Final report on the experimental and analysis programme, Internal AMEC-NNC report.

Tukey, J. W. (1977). Exploratory data analysis, Addison-Wesley.

Wallace, C., West, G. M., McArthur, S. D. J., Towle, D. and Reed, J. (2008). The effect of reactor parameters on AGR refuelling at Hinkley Point B. This volume – see pages 52-60.

West, G. M., Jahn, G. J., McArthur, S. D. J., and Reed, J. (2007), Graphite Core Condition Monitoring through Intelligent Analysis of Fuel Grab Load Trace Data. In: *Management of Ageing Processes in Graphite Reactor Cores (Ed. G. B. Neighbour)*, Proceedings of the Ageing Management of Graphite Reactor Cores held at the University of Cardiff on 28-30 November 2005, 232-239, RSC Publishing.

Volumetric Inspection of Reactor Core Graphite Using Eddy Currents

David Wood and Matthew Brown

David Wood from Serco, Rutherford House, Olympus Park Quedgeley Gloucester GL2 4NF
and Matthew Brown from British Energy, Barnett Way, Barnwood, Gloucester GL4 3RS
Email: david.wood@sercoassurance.com and matthew.brown@british-energy.com

Abstract
Current video surveys of channel bores in Magnox and AGR graphite reactor cores do not address potential concerns about subsurface cracking from the keyway slots. Development work carried out by the CEGB and Nuclear Electric in the late 80s and early 90s demonstrated that eddy currents could provide an inspection system capable of detecting cracks coming from the keyway slots. Work carried out by British Energy between 1999 and 2001 has confirmed that conclusion and extended it to include the detection of channel bore cracks. The development has continued from 2006 onwards as a collaborative programme jointly funded by British Energy and Magnox North. Initial development work has been carried out using virgin graphite. To provide confidence that the technique will work on irradiated graphite a finite element model has been developed to predict the responses of cracks in irradiated graphite. Additionally, some work has been carried out using low density filter graphite as a substitute for irradiated graphite. The development has now reached the stage where BE intend to develop a "Proof of Principle" inspection tool for in-core trials. Eddy current inspections rely on using a coil to induce eddy currents in a conductive material. As the coil is moved over the test surface, changes in the impedance of the coil are monitored. The standard depth of penetration of the eddy currents into a test material is defined as the depth at which the eddy current intensity drops to $1/e$ (~37%). It is a function of the test frequency, the magnetic permeability of the material and electrical conductivity. In the case of virgin graphite a test frequency of 1100Hz gives a depth of penetration, which is comparable with the depth of the keyways. Recent work on test blocks has concluded that slots with 39% through-wall penetration can be clearly detected. Some of the earlier work suggests that defects of smaller penetration can be detected and that a capability for sizing defects might be possible. The potential of eddy currents to measure and map graphite densities has also been considered.

Keywords
Eddy currents, Graphite, Cracks

INTRODUCTION

Currently Non-Destructive Testing on the graphite core of Magnox and AGR Power Stations is limited to video and dimensional surveys of graphite channels. The video survey provides a capability for detecting a variety of surface damage and cracks. However, it does not address potential concerns about cracking from the keyway slots, which are located on the outer surface of the bricks and are not visible.

On the basis of work carried out by British Energy over the last few years and earlier work carried out by the CEGB / Nuclear Electric, it seems likely that eddy currents can provide an in-reactor inspection technique for detection of cracking in graphite bricks. Of the various development programmes, the British Energy work, which was aimed at detection of defects in AGR bricks, was the most comprehensive and covered detection of cracking from keyway corners and channel bores. The issues on AGR bricks are sufficiently similar for this work to be equally relevant to Magnox bricks and the development continued from 2006 onwards as a collaborative programme jointly funded by British Energy and Magnox North.

GRAPHITE BRICK DESIGNS

Graphite Bricks are used as neutron moderators in the reactor core of all Magnox and AGR reactors. They lock together and form a matrix of channels, in which fuel and control rods lie. There are two basic brick designs of Magnox bricks: square and octagonal. Each brick is locked to adjacent bricks using graphite keys. Magnox channel bores are typically around 98mm in diameter. AGR graphite fuel channel bricks are circular with keyways located radially around the bricks. Bricks are keyed together in a similar way to Magnox bricks, using a combination of smaller filler bricks and control channel bricks to complete the matrix. Channel diameters are much larger than for Magnox reactors and typically of the order of 260mm in diameter. Additionally, there are approximately 80 small bore 'methane' holes running down the length of each brick.

PREVIOUS WORK

Development of non-destructive testing techniques for graphite reactor cores can be divided into three categories.

Early Development Work
Carried out by the CEGB and Nuclear Electric on the inspection of AGR graphite. This comprised firstly of work from the late 1980s carried out at the CEGB Marchwood Engineering Laboratories. There, test block trials were carried out to determine the capabilities of eddy current techniques for detection of keyway cracks. Secondly, work from the early 1990s carried out at the Nuclear Electric NDT Applications Centre. Both pieces of work suggested that eddy currents could detect cracks in AGR core bricks.

British Energy Development Programme
Work funded by British Energy between 1999 and 2001 and carried out jointly by Phoenix Inspection Systems and AMEC. This work started by reviewing previous work on electromagnetic techniques and considering what further work on electromagnetic techniques could be carried out. Ultrasonic techniques were considered, but not pursued, probably because of inherent difficulties associated with material attenuation and probe coupling. Work continued with eddy currents and used test blocks to establish a technique capable of detecting partial through-wall cracks and measuring their depth. The final phase of the work concentrated on the application of eddy currents to more realistic defects and the optimisation of test coils, equipment and procedures.

Windscale Pile 1
A desktop study carried out for BNFL in 2003 which considered a wide range of techniques for inspecting the whole core of Windscale Pile 1. The report from this study concluded that eddy currents were one of the most promising techniques for examination of graphite cores.

COMPARING AGR AND MAGNOX GRAPHITE BRICK DESIGNS

The inspection of graphite bricks for keyway cracks and cracks on the channel bore can only be carried out by inserting an inspection head into the bore of each channel. For detection of cracks running from keyways, the depth of the keyway below the channel bore is important. Typically AGR keyways are around 54mm below the channel bore surface, whereas Magnox brick edge keyways vary between 17mm and 43mm below the channel bore surface and for keyways at the corners of the octagonal bricks the depth varies between 65mm and 73mm. Figure 1 shows typical brick designs and Figure 2 brick layouts.

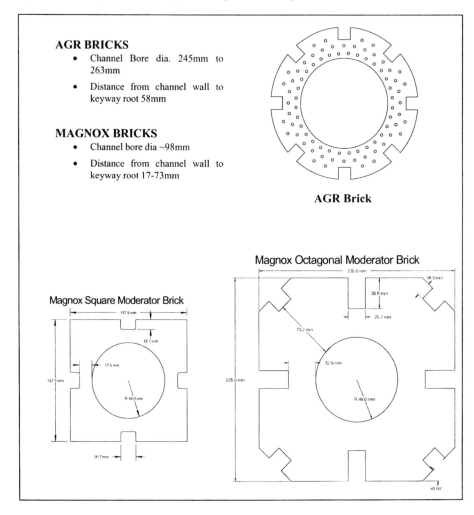

FIGURE 1: Graphite brick designs in AGR and Magnox reactors.

COMPARING AGR AND MAGNOX GRAPHITE MATERIAL

Magnox reactor cores were manufactured from Pile Grade A (PGA) graphite. AGR reactor cores were manufactured from Gilsocarbon graphite. PGA graphite was manufactured using a type of graphite known as needle coke because of the needle like appearance of the grains after crushing. The basic shape of the graphite blocks was formed in an extrusion process. This method has the effect of aligning the grains in a direction parallel to the direction of extrusion. As a result the properties are anisotropic. Gilsocarbon graphite was produced to provide a more dimensionally stable graphite. It was produced from Gilsonite pitch coke, which has rounder grains, using a moulding process which ensures near isotropic properties.

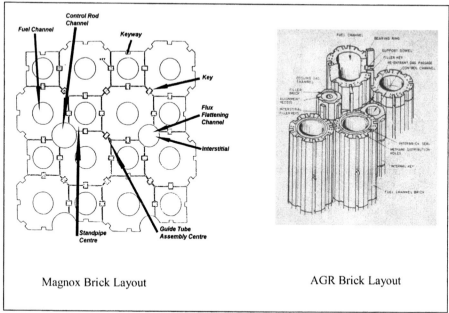

FIGURE 2: Brick layout in Magnox reactors and AGRs.

ULTRASONIC TECHNIQUES

Graphite is ultrasonically a very attenuative material and the use of ultrasonics to measure Young's modulus on trepanned core samples has shown that this is particularly the case with high weight loss graphite.

British Energy work on ultrasonics stopped after the initial study, presumably because of difficulties associated with coupling the probes to the test surface and the high attenuation of the graphite. The conventional method for coupling ultrasonic probes, uses a liquid or viscous gel between the probe shoe and test surface. There are alternative means of generating ultrasound within a test material such as electromagnetic acoustic transducers (EMAT probes) and laser techniques. Both EMAT and laser ultrasound systems have been around for a number of years. However, neither technique has been developed to a state where it is routinely used and accepted. The feasibility of using either technique with graphite is unknown. Inspection speeds are also a consideration. It seems likely that, given the need to inspect long lengths of graphite channel, inspections using an ultrasonic system would be relatively slow to carry out.

EDDY CURRENTS

A programme of eddy current development work was funded by British Energy between 1999 and 2001. This continued as a jointly funded British Energy / Magnox North programme from 2006 onwards, with the aim of resolving a number of outstanding technical issues relating to the capabilities of the eddy current technique.

An eddy current inspection system for graphite offers the prospect of a fast non-contact technique. As part of the development programme a range of eddy current parameters were considered. Probe designs covering axial (266mm diameter) and radial (24mm and 70mm diameter) coils were assessed. Frequencies from 1kHz to 60kHz were applied in multi-frequency mixed modes to cancel unwanted sources of interference such as lift off and conductivity variations. Eddy currents were applied to detect simulated defects of varying depths. This included saw cut defects and artificially induced cracks running from the channel bore and from the corners of keyways. The use of phase and amplitude to measure defect size and depth was evaluated. The effect of compression on the crack detection reliability was assessed.

Probe Design
An axial probe with a diameter of 266mm was tested. This was found unsuitable for detecting compressed cracks, although it would easily detect a 100% through-wall defect where no material contact existed between the faces. The reason for its poor performance on compressed cracks was put down to the large detection area in relation to the low amplitude and extent of electric field disturbances. Subsequent work concentrated on radial probes employing 70mm diameter coils for detection of deeper defects and 24mm coils for detection of shallower defects.

Eddy Currents and AC Skin Effect
The penetration of eddy currents into a component is a function of the AC skin effect. The following standard formula is used:

$$\delta = \frac{1}{\sqrt{\pi \upsilon \mu \sigma}}$$

where :

δ = standard depth of penetration   [1]

υ = frequency

μ = magnetic permeability (assumed to be that of free space $(4.\pi.10^{-7}$ H.m$^{-1})$

σ = electrical conductivity

This indicates that at a frequency of 1100Hz, using a resistivity value of 1006μΩ.cm (from manufacturers heat certificates), which is equivalent to a conductivity of 99,400 Sm^{-1}, the standard depth of penetration will be 48mm. The standard depth of penetration is defined as the depth at which the eddy current density has decreased to 1/e, or about 37% of the surface density.

Eddy Current Coil Size
The larger the coil the greater the coverage. However, inspecting large areas with large coils will result in reduced inspection sensitivity. This is illustrated by the work carried out for British Energy where graphite channel test pieces were inspected using a single axially wound coil with a diameter of 266mm. This approach was abandoned because of its poor sensitivity to keyway corner and tight cracks. However, larger coils tend to perform better at low frequencies and thus provide better depth of penetration. In this work 70mm coils were

used at frequencies down to 2kHz, in which mode the keyways were detected. Coils of 24mm diameter were used at frequencies down to 3.5kHz to detect and assess near surface defect responses.

Multi-Frequency Techniques

Multi-frequency techniques are used in eddy current testing to eliminate the effect of unwanted influences. At different frequencies the phase angle for responses will change. Under such circumstances the mixing of responses will enable unwanted effects to be eliminated. For the 70mm diameter coil it was found that a combination of 2 and 10kHz frequencies were successful in cancelling conductivity variations and reducing lift-off variations. For the 24mm coils, a range of dual frequency set-ups were used (4.5 and 18.5kHz, 4.5 and 55kHz, 9 and 55kHz). The largest unwanted response came from lift-off, and frequency mixing was used to cancel this effect. It was found that a digital band pass filter, enabled from within the display software, could be used to minimise the effect of conductivity variations.

Detection Capability

Saw cuts of increasing depth were used to calibrate the eddy current system. This was done by relating saw cut depth to amplitude and phase of response. For keyway saw cuts this was recorded for a 70mm coil operating in dual frequency 2/10kHz mode and in 2kHz single frequency mode. A response was recorded for remaining ligaments of up to 50mm in 2/10kHz multi-frequency mode and up to 30mm in 2kHz single frequency mode.

One of the development objectives was to investigate the capability of the eddy current technique to detect tight cracks. Crack detectability under compression was measured by applying compressive loads to crack faces and recording the change in response. Loads of up to 1000kg were applied and this reduced the response in some cases by up to 82%. This demonstrates the susceptibility of the response to compressive forces.

To demonstrate the capability for inspecting whole bricks, simulated defects were introduced into complete bricks and scanned with a computer controlled manipulator, using a range of setups. Figure 3 shows the results of a helical scan of one of those test blocks. From this work it was concluded that a 39% through-wall slot cut at a keyway corner could be reliably detected. This equates to a defect with a through-wall size of 23mm and a crack tip depth from the channel bore of 35mm.

Sizing Capability

Defects can potentially be sized using either amplitude or phase angle of response. The effect of compressive load on response suggests that amplitude is not likely to be a particularly reliable method for measuring crack size. Phase angle was found to be a more accurate method and less affected by compressive load.

Mathematical Modelling

A finite element model was developed to predict eddy current responses in unirradiated graphite. Measured responses from test pieces were used to validate the model. This model was then extended to predict the eddy current response in irradiated graphite. The model was able to predict the relative amplitude and direction of defect signal responses at selected operating points on the impedance plane curve. A limited amount of work using low density filter graphite has provided some support for the predictions. However, assumptions on the electrical conductivity and magnetic permeability of irradiated graphite have had to be made and the actual inspection capability for both Magnox and AGR reactors will remain unknown until in-reactor trials can be carried out.

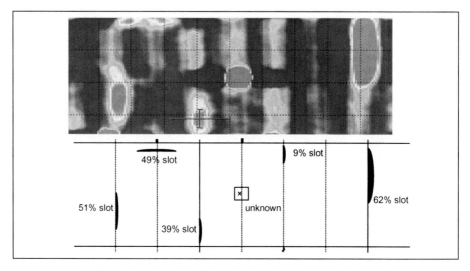

FIGURE 3: Helical Scan of Test Block Red10 with Diagram of Defects and
Eddy Current Response (4/20kHz).

Scanning Considerations
The capabilities of an array of coils compared with the more usual single coil rotating head arrangement have been assessed. This concluded that for 70mm diameter coils on unirradiated graphite an array of coils with a 15° circumferential spacing in an AGR fuel channel (~260mm diameter) gave a performance equivalent to a helical (rotating head) scan. Twenty four coils in two circumferential rows would provide 360° coverage. This equates to a 40° circumferential spacing with 9 coils for the smaller (~100mm diameter) Magnox fuel channels. Multiplexing of the coils and simultaneous lowering of the array would enable a fuel channel to be fully inspected at a rate of 25 to 50mm per second.

CONCLUSIONS

- For unirradiated AGR bricks, a slot cut at a keyway root that penetrates 39% through the keyway to bore thickness was clearly detected.
- A fixed array of coils could be used as an alternative to a rotating head and should provide a capability for inspecting fuel channels at a rate of 25 to 50mm per second.
- A mathematical model has been used to predict the capability of eddy currents on irradiated graphite. To do this some assumptions have to be made on the electrical conductivity and magnetic permeability of irradiated graphite. The actual capability for both Magnox and AGR reactors will not be known until in-reactor trials can be carried out.

Acknowledgements
The authors would like to thank British Energy Generation Ltd, Magnox North and Serco for permission to present this paper. The recent joint funded British Energy / Magnox North development programme was managed by Amec NNC and carried out by J. L. Turner – Phoenix Inspection Systems, T. J. Bloodworth – Independent Consultant and C. Holt – ESR Technology.

A Statistical Analysis of the Mechanical Properties of PGA Graphite Samples Taken from Magnox Nuclear Reactors

Kevin McNally[a], Emma Tan[a], Nick Warren[a], Graham B. Heys[b] and Barry J. Marsden[c]

[a]Mathematical Sciences Unit, Health and Safety Laboratory, Harpur Hill, Buxton, UK
[b]Nuclear Installations Inspectorate, Redgrave Court, Bootle, UK
[c]Nuclear Graphite Research Group, School of Mechanical, Aerospace and Civil Engineering, University of Manchester
Email: kevin.mcnally@hsl.gov.uk

Abstract
Results of statistical analyses of strength and dynamic Young's modulus (DYM) data are presented for installed set and trepanned graphite samples taken from various Magnox reactors. Also DYM data from graphite samples irradiated in a recent Material Test Reactor (MTR) programme are presented. In reactor, the properties of these samples were modified by fast neutron irradiation and radiolytic weight loss. Both conventional decay and threshold models were fitted to these data as a function of weight loss. The strength data showed evidence of threshold limit in strength, whereas the DYM data did not. At low fast neutron dose, both strength and DYM appear to be influenced by the ratio between fast neutron damage and weight loss. The Brazilian disc test is considered to be unsuitable for samples with high weight loss due to localised crushing at the load points.

Keywords
Strength, Dynamic Young's modulus, Radiolytic weight loss, PGA graphite

INTRODUCTION

The moderator used in the UK fleet of Magnox reactors is Pile Grade A (PGA) graphite. PGA is a porous material and key mechanical properties, such as strength and Young's modulus, are known to decrease as the volumetric porosity of the material increases. During reactor irradiation, the strength and modulus of PGA change due to the competing processes of, fast neutron irradiation, which causes crystallite hardening, and by radiolytic oxidation, which increases porosity. The combined effect of these two processes is assumed to be multiplicative, referred to as the 'product rule' (Metcalfe and Payne, 2007).

The experimental work of Ryshkewitch (1953) showed that the compressive strength of two ceramics (alumina and zirconia) varied logarithmically with a linear change in porosity. Based on a review of the available literature, Knudsen (1959) found that a relationship of this form, referred to in the graphite literature as the Knudsen model, was appropriate for other ceramic materials and for compressive, tensile and flexural strength. The Knudsen model has since been applied to other mechanical properties. The mechanical properties of PGA irradiated in Materials Test Reactors (MTR) have been modelled using a similar exponential decay model (Kelly et al., 1983), but with weight loss used in place of porosity. The Knudsen model has been found to fit data for porosities of up to 50% (Rice, 1998), a value that corresponds to a weight loss of approximately 40% in irradiated PGA. In comparison, the PGA MTR data were sparse beyond 20% weight loss and terminated at a weight loss of approximately 35%. In reality there must be a physical limit at which the porosity increase is so great that strength and modulus decrease to zero; for this reason, two alternative models that are suitable for materials with high levels of porosity were presented in Rice (1998). These models are referred to as threshold models on account of a threshold term that dictates a bound at which a given material property reaches zero.

This paper presents results from statistical models, based upon an exponential decay and two threshold models (Rice, 1998). The models were fitted to strength and Young's modulus data measured on samples taken from six Magnox power stations. These materials properties data were measured on 'installed sets', which are PGA samples loaded in interstitial channels for core surveillance, and trepanned cores sampled from PGA moderator bricks. Young's modulus data were also available from samples taken from two unirradiated PGA graphite blocks and irradiated in carbon dioxide at the Idaho National Engineering and Environmental Laboratory (INEEL) in a recent MTR experiment.

METHODOLOGY

Mechanical Property Test Procedures

Strength
The strength of the installed samples was measured in compression on cylinders, in uniaxial tension on dumbbell type specimens, and in bending on four point flexural specimens. The tensile strength of the trepanned samples was only measured using the Brazilian Disc Method (Li *et al.*, 2007). A number of installed-set samples were also tested by the Brazilian disc method to determine tensile strength. This technique involves the compression between two anvils of 12 mm diameter (D) x 6 mm thick (t) discs cut from the trepanned specimens. The maximum tensile stress at the centre of the disc (σ) is calculated from the failure load (P) as:

$$\sigma = \frac{2P}{\pi Dt}.$$

[1]

Equation [1] is for an isotropic material so correction factors of 1.09 and 1.30 were used to account for the perpendicular and parallel directions respectively to determine the uniaxial tensile strength, these factors were derived by comparing results of disc compression tests with conventional tensile tests. The design of the anvil includes a 9mm radius to avoid as far as possible local failure by crushing at these points instead of in the centre of the disc.

Dynamic Young's Modulus (DYM)
The DYM measurements were made using transducers with a frequency of ~0.5 MHz. The measurement technique is based on ASTM C769 – 98 (ASTM, 2005) and the DYM (E) is determined using the relationship:

$$E = \frac{\rho v^2}{1.02}$$

[2]

where E is Young's modulus (Pa), ρ = density (kg/m^3), v = signal velocity (m/s) and 1.02 is a function of the Poisson's ratio for PGA graphite. For use in structural integrity calculations, DYM needs to be converted to static Young's modulus using an appropriate factor. There were three developments of the apparatus used to determine DYM - Mk1, Mk2 and Mk3.

Mechanical Property Data

Installed-set samples
Installed-set samples of PGA graphite are of varying shapes and sizes loaded in the interstitial channels of the reactors at start-of-life. The purpose of installing such samples was to enable monitoring of chemical and physical properties without need for invasive techniques. As fast neutron doses within interstitial channels are lower than within fuel channels, titanium tie-rods were used to enhance the energy deposited (GTAC, 2005) in the graphite samples resulting in increased radiolytic weight loss. As a result, installed-set samples with the highest weight losses lead losses in moderator bricks. Recent carrier withdrawals and testing of installed-set samples has produced a large dataset of mechanical property test results. A small proportion of installed-set samples were machined into smaller specimens before testing.

Table 1 summarises strength and DYM data measured on installed-set samples. An additional dataset of DYM measurements, made on samples machined from two blocks of reactor graphite and irradiated at the INEEL MTR facility, supplemented the data available from commercial Magnox power stations.

TABLE 1: Summary details of the available mechanical properties data from installed sets.

Mechanical property (units)	N	Source	Weight loss (%)	Mechanical property
Uniaxial tension (MPa)	122	3 power stations	1.22 – 37.40	1.14 – 36.54
Tensile strength (MPa)	122	1 power station	5.50 – 13.00	5.02 – 14.69
Flexural strength (MPa)	91	1 power station	13.0 – 45.15	0.55 – 22.70
Compressive strength (MPa)	349	5 power stations	0.71 – 47.90	0.71 – 66.80
DYM (GPa)	189	5 power stations	0.47 – 43.80	1.40 – 26.38
DYM (GPa)	40	INEEL	18.8 – 57.95	0.59 – 12.69

Trepanned cores
Graphite samples are routinely trepanned from Magnox reactor fuel and interstitial channels during reactor core inspections. The trepanned cores are cylindrical in shape with a circular cross section and dimensions of approximately 12 mm diameter and 20 mm depth. The cores are typically machined into three slices of approximately 12 mm in diameter and 6 mm in depth before mechanical testing; slice one is from the channel wall surface. Due to the difficulty in conventionally measuring strength on samples of this size and geometry, the tensile strength data are all measured using the Brazilian disc test. The available data on strength and DYM are summarised in Table 2.

TABLE 2: Summary details of the available mechanical properties data from trepanned cores

Mechanical property (units)	N	Source	Weight loss (%)	Mechanical property
Tensile strength (MPa)	476	6 power stations	0.00 – 32.80	1.53 – 21.79
DYM (GPa)	1030	6 power stations	0.80 – 39.30	1.70 – 18.20

Statistical Models
Statistical models based upon an exponential decay function were fitted to each of the mechanical properties in Tables 1 and 2. The initial model was of the form:

$$Y = Y_0 \exp(-\beta w).$$ [3]

where Y represents the measured mechanical property and w represents the corresponding percentage weight loss. The model parameters are Y_0, which is the expected value of the mechanical property for unirradiated material, and, β, the decay rate. Models were specified on the log-scale using a mixed-effect analysis. Additional statistical parameters are the random effects, which quantify the systematic difference between the trend line and measurements made on the same sample (for example, measurements made on three slices of a trepanned core would share a common random effect). For most installed sets only a single measurement was made, however, a small number were pre-machined into smaller samples before testing, hence multiple measurements from a common installed set were available. The majority of trepanned cores had three measurements available for both tensile strength and DYM. Random effects and residuals, which represent sources of variability in the statistical model, are assumed normally distributed with zero means and standard deviations of ψ and σ respectively. An assumption of normality on the log-scale implies asymmetry about the trend line and variability proportional to magnitude on the data-scale.

Additional terms, allowing for a change to the trend by incorporating powers of weight loss, and for differences between orientations (relative to sample extrusion direction), channel (fuel or interstitial), heat treatment, power station, and measuring equipment (Mk1, Mk2 and Mk3 for DYM) were added to the simple model sequentially, based upon an analysis of model residuals. A more elaborate variance structure was also introduced when the model residuals indicated this was appropriate. For each model, rigorous diagnostic testing was undertaken to investigate goodness of fit and to ensure no systematic trends in model residuals.

Similarly, the two threshold models referred to in Rice (1998) as the Stress Concentration Effects (SCE) model, originally proposed by Phani and Niyogi (1987), and Schillers equation (Schiller, 1960) were fitted. The initial models were of the respective forms [4] and [5].

$$Y = Y_0 (1 - \frac{w}{T})^n$$ [4]

$$Y = Y_0 [1 - (\frac{w}{T})^n]$$ [5]

In [4] and [5] parameter n represents decay and T a threshold weight loss, beyond which the mechanical property is zero. The remaining terms, common to Equation [1], have the same interpretation as previously given. The models were fitted and extended using a non-linear mixed-effect analysis, based upon a rigorous analysis of model residuals.

RESULTS

Strength

Exponential decay
The intercept terms in the statistical models correspond to estimates of the geometric mean virgin strength. Estimates are consistent with the known order of test method, with compressive tests resulting in the highest strengths (45.9 MPa) followed by flexural (18.4 and 9.1 MPa for tests parallel and perpendicular to extrusion direction respectively), then tensile failure by Brazilian Disc test (13.2 MPa) and lastly uniaxial tension (10.8 and 5.3 MPa

respectively for tests parallel and perpendicular to extrusion direction). However, these estimates are not consistent with directly measured virgin strength data. This is probably because of the low dose counter balance between the rate of increase in strength due to fast neutron irradiation damage, and the rate of decrease in strength due to radiolytic oxidation. Statistically significant differences were found in the rates of decay between strength measures. The decay rate was greatest for flexural strength followed by compressive strength, tensile failure by Brazilian Disc test and lastly uniaxial tension. A model of the form Equation [3] fit the data on compressive and flexural strength beyond 40% weight loss poorly; however, the addition of higher order powers (cubic) of weight loss resulted in a good fit to the data at these larger weight losses. Data on the other strength tests were not available at such high weight losses (Table 1). Only data on compressive strength and uniaxial tension data were available for more than one power station and the overlap in weight loss was small (Table 1). There were differences between power stations for these two measures consistent with differences in the relationships between irradiation dose and weight loss. Samples from the power station showing the most rapid accumulation in dose relative to weight loss had the greatest strengths.

Only measurements of tensile strength, measured using the Brazilian Disc test, were available from trepanned cores. The intercept was consistent with that obtained for installed sets, however, the larger dataset allowed a modest but statistically significant difference between parallel and perpendicular orientations (14.8 and 12 MPa, respectively) to be determined. The trend line for the interstitial samples was similar to the tensile measurements (measured using the Brazilian Disc test) on installed sets over the weight loss range where data from both sources were available. Statistical parameters measuring variability about the trend line (the random effect and residual variance components) were also very similar. Interestingly, the tensile strength of the first slice for trepanned cores from fuel channels decayed at a slower rate than for slices two and three. The weight loss was also noticeably lower for the first slice of a fuel channel core. All slices from interstitial channels decayed at the same rate as slices two and three from fuel channels.

Thermal annealing of the samples prior to measurement reduced strength in both installed sets and trepanned cores significantly (approximately 60% and 40%, respectively). There was more variation in strength (relative to magnitude) noted at the two reactors showing the highest weight losses.

Threshold models
Two threshold models were fitted to data from installed sets and trepanned cores. These models were consistent with the exponential decay models, and provided a similar fit to the data up to the highest currently observed weight losses for flexural and compressive strength (Figure 1). Beyond the limit of the available data, the threshold models rapidly diverged from the exponential decay models. In particular, these models predicted thresholds of approximately 48% for flexural strength and 50-57% weight loss for compressive strength. For the Brazilian Disc test, a threshold was predicted at 32-35% and approximately 38% weight loss for parallel and perpendicular orientations, respectively.

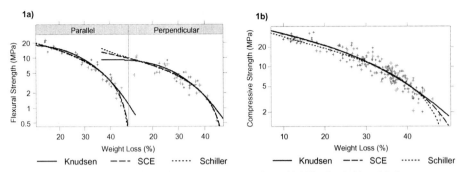

FIGURE 1: Plots of fitted exponential decay and SCE and Schiller threshold models for:
a) flexural strength with parallel and perpendicular orientations; b) compressive strength.
Plotted data are from the power station showing the highest weight losses

DYM

Exponential decay

Trepanned cores offered the richest source of information on DYM with many measurements made using both the Mk1 and Mk2 equipment, from multiple stations and an overlapping weight loss range during which the two test rigs were used. Due to the nature of the test, in all cases DYM was measured perpendicular to extrusion direction. The intercept for slices 2 and 3 using the Mk2 equipment was 7.5 GPa. Intercepts for slice 1 in fuel (FC) and interstitial channels (IC) were greater (9.4 and 7.9 GPa, respectively). Measurements of DYM using the Mk1 equipment were approximately 45% higher than those using Mk2. Differences between power stations were identified by the analysis; at one power station, with measurements made between 3.4% and 16.3% weight loss, DYM measurements were approximately 23% and 7% above the best estimate trend line for all fuel and interstitial channels respectively. A second power station, with 535 measurements available at weight losses between 0.8 and 39.3%, had DYM measurements approximately 11% and 18% below the best estimate trend line for slice 1, and slices 2 and 3 respectively. All the differences related to the intercepts borne out by the power station-specific trend lines being parallel to one another. The model was significantly improved by including a quadratic term for weight loss; such models have been applied to Young's modulus data previously (Yoshimura *et al.*, 2007). The important decay curves can be seen in Figure 2a.

Measurements were available on installed sets from four power stations at weight losses between 0.5% and 16.3% using the Mk1 equipment, and from one power station at weight losses between 26.5% and 43.8% using the Mk2 equipment. The data were sufficient to show that decay curves were of a similar form to installed sets (Figure 2b), and that DYM measured parallel to extrusion direction was approximately 30% greater than perpendicular to extrusion. No difference between Mk1 and Mk2 was found. The Mk1 data suggested there were various differences between power stations however there were insufficient data to quantify all the differences with confidence. There was more variation about the trend line at the two reactors showing the highest weight losses for both installed sets and trepanned cores.

Graphite samples from the two virgin PGA blocks irradiated at the INEEL using Mk3 equipment, which tests have shown provide similar measurements to the Mk2 equipment (Tzelepi *et al.*, 2007*)*, were the only data that showed differences in the decay rates. It is unclear why the blocks originally intended for different power stations behaved differently;

however, since each batch of samples were from a single brick, manufacturing variability could be one explanation. The decay curve for samples from one of the bricks was similar to the decay curves for the power station for which the block was intended (Figure 2b).

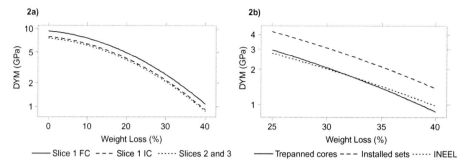

FIGURE 2: Plots of fitted exponential decay curves to DYM data highlighting:
a) key differences between slice 1 and slices 2 and 3;
b) a comparison of trend curves for the three data sources available at one power station.

Threshold models
Two threshold models were fitted to these DYM data. The models identified a similar structure to the exponential decay models; however the thresholds were estimated as 100%. This material property will decay to zero before 100%, however at present even the data from the INEEL MTR experiment did not extend to high enough weight losses to suggest a threshold is being approached for this property.

DISCUSSION

The poor fits of both strength and DYM models to materials at low dose, along with the inaccurate prediction of virgin properties are most probably related to differences in rate of dose accumulation relative to rate of weight loss increase. There will be significant differences in the ratio of these rates between the INEEL, installed and trepanned samples, and some differences in this ratio between reactors.

For compressive strength, the estimated threshold is around 50% to 60% weight loss, depending on the model chosen, and 48% for flexural strength under both threshold models. The threshold is sensitive to the high weight loss data and, although there is an apparent lower threshold for disc compression specimens (around 30% to 35% weight loss), it cannot be certain that this is not an artefact of the test itself as tensile failure by diametral compression can only be applied up to medium weight loss, due to the possibility of crushing at the load point. The disc compression strength data do not extend to the higher weight losses tested by compression.

It has been postulated that for graphite, there is a power law relationship between modulus and strength. To explore this relationship for irradiated and radiolytically oxidised graphite, DYM and strength measurements would be required on the same samples; unfortunately these are not available. However, it is clear from the data presented that both these properties can be described using similar models related by some form of power law, up to the point where strength starts to decrease at a faster rate as a threshold limit is reached, whereas DYM even at high weight loss does not exhibit a threshold. This may indicate there is a change in failure mechanism at high weight loss, such as from a fracture mechanism to ligament collapse.

CONCLUSIONS

1. There are differences in strength decay curves for installed sets from different power stations, however no differences were found between stations in decay curves for trepanned cores.

2. The installed, trepanned and INEEL samples for both strength and modulus followed similar decay curves. However, the virgin properties were not fitted well by the models. It is suggested that this reflects the difference in the ratio of fast neutron damage to radiolytic oxidation rate for the different samples.

3. Evidence of a threshold limit was found in all the strength data. The weight loss at which the threshold occurred appears to be a function of the type of test. In particular the Brazilian disc test is considered unsuitable for testing high weight loss samples due to local crushing at the load points, resulting in the prediction of a low threshold.

4. Dynamic Young's modulus did not exhibit a threshold limit, whereas strength did. This may indicate a change of failure mechanism at high weight loss.

Acknowledgements
The authors wish to thank Magnox Electric for permission to use mechanical property data measured on PGA graphite samples removed from Magnox power stations. The views expressed in this paper are those of the authors and do not necessarily represent the views of HSE or HM Nuclear Installations Inspectorate.

References
ASTM C769 - 98(2005). Standard test method for sonic velocity in manufactured carbon and graphite materials for use in obtaining an approximate Young's modulus.

Li, H., Fok, A. S. L. and Marsden, B. (2007). Evaluation of the Brazilian disc test in fracture strength measurements of nuclear graphite. In:- *Management of Ageing Processes in Graphite Reactor Cores* (Ed. Gareth B. Neighbour), 75-82, RSC Publishing, ISBN 978-0-85404-345-3.

Melcalfe, M. P. and Payne, J. F. B. (2007). The effects of thermal annealing on the mechanical properties of PGA graphite. In:- *Management of Ageing Processes in Graphite Reactor Cores* (Ed. Gareth B. Neighbour), 35-42, RSC Publishing, ISBN 978-0-85404-345-3.

NII Graphite Technical Advisory Committee (GTAC) for Nuclear Plant (2005). Radiation dosimetry: calculation of the spatial and temporal variation of energy deposition in graphite bricks.

Kelly, B. T., Johnson, P. A. V., Schofield, P., Brocklehurst, J. E. and Birch, M. (1983). U.K.A.E.A. Northern Division studies of the radiolytic oxidation of graphite in carbon dioxide. *Carbon*, 21, [4], 441-449.

Knudsen, F. (1959). Dependence of mechanical strength of brittle polycrystalline specimens on porosity and grain size. *Journal of the American Ceramic Society*, 42, [8], 376-387.

Phani, K. K. and Niyogi, S. K. (1987). Young's modulus of porous brittle solids. *Journal of Material Science*, 22, [12], 257-263.

Rice, R. (1998). *Porosity of Ceramics*. Taylor and Francis CRC press.

Ryshkewitch, E. (1953). Compression strength of porous sintered alumina and zirconia - 9th communication to ceramography. *Journal of the American Ceramic Society*, 36, [2], 65-68.

Schiller, K. K. (1960). Skeleton strength and critical porosity in set sulphate plasters. *British Journal of Applied Physics*, 11, [8], 338-342.

Tzelepi, N., Brown, M., Matthews, P. C., Payne, J. F. B. and Metcalfe, M. P. (2007). The development of measurement techniques for mechanical properties applicable to small reactor graphite samples. In:- *Management of Ageing Processes in Graphite Reactor Cores* (Ed. Gareth B. Neighbour), 142-149, RSC Publishing, ISBN 978-0-85404-345-3.

Yoshimura, H. N., Molisani, A. L., Narita, N. E, Cesar, P. F. and Goldenstein, H. (2007). Porosity dependence of elastic constants in aluminium nitride ceramics. *Materials Research*, 10, [2], 127-133.

Understanding AGR Graphite Brick Cracking Using Physical Understanding and Statistical Modelling

Philip R. Maul, Peter R. Robinson and Alan G. Steer[#]

Quintessa Limited, The Hub, 14 Station Road, Henley-on-Thames, UK
[#]Design Authority, British Energy, Barnett Way, Barnwood, Gloucester, UK
Email: philipmaul@quintessa.org

Abstract

Most bore-initiated cracks in graphite bricks cracks are caused by the creation of a defect of critical size and shape when the graphite is subjected to high stress or strain. Our understanding of the development of these cracks has greatly improved in recent years due to a combination of advances in stress analysis, core monitoring and statistical modelling. This paper summarises progress in these areas and shows how consistently good 'blind' predictions of crack numbers observed at core inspections have been made. Physically-based models suggest that the peak risk of cracking occurs between 10 and 20 full power reactor years, and this is consistent with evidence from several sources that cracking rates have reduced, possibly to zero, at the lead stations.

Keywords

AGR, Graphite properties, Statistical models

INTRODUCTION

The monitoring of the physical integrity of graphite cores in Advanced Gas-cooled Reactors (AGRs) is an important part of the inspection leg to the safety case for their continued operation. Maul and Robinson (2007) gave details of a strategy for the monitoring of bore cracks using statistical techniques. The approach provides a transparent and auditable justification for the safe operation of the reactors until the next inspection. In the last three years much more data have become available on the statistics of bore cracks and a much improved understanding of the causes of these cracks has been obtained.

The paper is structured as follows:

- A brief summary of the statistical modelling methods is given.
- Examples of model predictions are presented to illustrate the power of methodology.
- A summary of the physical understanding of the processes involved is given.
- Finally, the main conclusions are drawn together and a discussion of developments that can be anticipated in the next few years is presented.

STATISTICAL MODELLING TECHNIQUES

Monitoring of the graphite core involves TV inspections, Channel Bore Measurement Units (CBMU), and Trepanning. No bore cracks were observed prior to 1994, but since then cracks of various morphologies have been observed, most have been found in fuel bricks of the upper moderator layers. Full-length axial cracks are considered to have the greatest potential to affect the functionality of the radial keying system. Since this was designed to maintain the core channels straight and vertical, the statistical models have been applied to such cracks. The methodology can, however, be applied to any class of bore cracks and results for circumferential cracking are referred to here.

Bricks in the same layer and in the central moderator layers broadly experience the same operating conditions and so are generally assumed to be statistically identical, although calculations have been undertaken based on the assumption that different types of brick have different cracking risks (*e.g.* those in 'edge' channels have significantly different dose profiles to those in 'central' channels).

An important principle is that the approach must be capable of extrapolation; the aim is not to find a single 'best' model, but to delimit an envelope of models that are consistent with the available data. These models will behave differently when extrapolated: this helps in the quantification of the level of confidence that can be associated with in forward predictions.

The methodology is iterative, as each new inspection provides additional data. A key feature is the ability to compare predictions made before the inspection with actual observations; this gives confidence in the use of the models, and enables models that cease to be consistent with the data to be rejected.

Full mathematical details will not be given here; the key concepts are described in Maul and Robinson (2006) and (2007). The probabilities per unit time of a brick changing from being uncracked (state 0) to singly cracked (state 1) and from singly cracked to doubly cracked (state 2) are defined in terms of hazard functions expressed in the following simple forms:

$$h_0(t) = u\,f(t); \quad h_1(t) = v\,f(t). \tag{1}$$

Here t can be any suitable age measure for the brick; calendar date has been used, although the most recent studies have shown a slightly better fit to the data if core burn up is employed. The forms that have been used to date for the function $f(t)$ can be written in terms of the following general expression:

$$f(t) = \max\left[0, \sum_{k=1}^{K} a_k\,(t-t_0)^{p_k}\right] H(t-t_0), \tag{2}$$

where H is the Heaviside step function and t_0 is the time when the hazard is assumed to commence. In practice, the number of terms K in the summation will be small: only Equations [1] or [2] have been used in the models employed to date. With a single term the two simplest two-parameter models considered are: a Uniform Model, where the hazard is constant ($p=0$) after a specified time, and a Linear Model, where the hazard increases linearly ($p=1$) after a specified time. The cracking hazard may decrease with time if a negative value of p is used or if two terms are used in the summation, one of which is negative; a Quadratic Model is obtained if a_1 is taken to be unity and a_2 is taken to be a negative free parameter with $p_1 = 1$ and $p_2 = 2$.

An important consideration in fitting the free model parameters to the data is that the data are 'censored' in that an observation of a crack at a given time/age only indicates that the crack occurred sometime previously; it does not provide information on precisely when the crack occurred.

EXAMPLE CALCULATIONS

The lead AGR stations in terms of brick cracking rates are considered in two groups: Hinkley Point B (HPB) and Hunterston B (HNB); and Heysham 1 (HYA) and Hartlepool (HRA). Generally data from each group are pooled, but calculations can be undertaken for individual stations or reactors.

An example calculation is summarised here for the inspection that was undertaken at HYA R2 in May 2008. For this inspection the following statistical models were employed for full-length axial cracks:

1. A Linear Model (2 free parameters), with a linearly increasing hazard, starting from zero at a reference origin;
2. A Power Model (3 free parameters), with a hazard increasing with a general power of time, starting from zero at a reference origin;
3. A Delayed Uniform Model (3 free parameters) which allows for a delay time before a constant hazard is used;
4. A Delayed Linear Model (3 free parameters) which allows for a delay time before a linearly increasing hazard is used;
5. A General Model (4 free parameters), which allows for a delay time before the hazard starts and for any power of time;
6. A Delayed Uniform Mixed Risk Model (4 free parameters) which is the same as the Delayed Uniform Model with the addition of a multiplicative risk factor that is applied for edge channels; and,
7. A Delayed Quadratic Model (4 free parameters) which allows for the possibility that the hazard may start to fall at some time.

One criterion for determining how well a model fits the available data is the AIC (see, for example, Davison, 2003); this criterion trades off how close the calculations are to the measurements against model complexity. All the models considered are consistent with the available data, but the Delayed Uniform and Delayed Quadratic models performed best. This was the first time that a four-parameter model was considered to be the best on the basis of the AIC; previously less complex models were favoured. This suggests that the currently available data are now adequate to support slightly more complex models. The maximum likelihood parameter values for the Delayed Quadratic Model suggest that the risk of bore cracking may actually have already ceased. The maximum likelihood estimator for the parameter p in the Delayed Quadratic model is negative, which is also consistent with current risks of cracking being less than levels applicable earlier in the reactors' histories.

Figure 1 shows the pre-inspection predictions for the outage at HYA R2 in May 2008: the number of new singly-cracked bricks is shown on the vertical axis and the number of new doubly-cracked bricks on the horizontal axis. The area in the figure to the edge of Region 1 indicates where the observations are expected to be 50% of the time, the area to the edge of Region 2 95% and the area to the edge of Region 3 99% of the time. The actual observed number of cracks (1 new double and 2 new singles) is indicated by the black square in the figure.

	D0	D1	D2	D3	D4	D5
S0	2	2	3	4	4	4
S1	2	2	2	3	3	4
S2	1	2	2	2	3	4
S3	1	1	2	2	3	4
S4	1	1	2	2	3	4
S5	1	2	2	2	3	4
S6	1	2	2	3	3	4
S7	2	2	2	3	4	4
S8	2	2	3	3	4	4
S9	2	2	3	4	4	4
S10	3	3	3	4	4	4
S11	3	3	4	4	4	4
S12	3	4	4	4	4	4

FIGURE 1: Pre-inspection predictions for HYA R2 May 2008 for the Delayed Quadratic Model.

Table 1 summarises the calculations undertaken for the HYA/HRA group of reactors in the last four years. The predicted and observed numbers of single and double cracks for the best statistical model (on the AIC basis), are compared, and an approximate range is given for the observations that would be deemed to be compatible with the model predictions. Over the eight inspections the numbers of observed singly and doubly-cracked bricks have always been consistent with the predictions and the best estimate total of 31 single cracks predicted compares well with the 27 observed.

TABLE 1: Recent inspections at HYA/HRA with the Linear Model and full-length axial cracks.

Reactor	Date	Best Model*	Single Cracks Predicted	Single Cracks Observed	Double Cracks Predicted	Double cracks Observed
HYA R2	May 05	L	2 (0-8)	8	0 (0-5)	0
HYA R1	Sep 05	DL	6 (0-11)	3	0 (0-4)	1
HRA R1	Jan 06	L	4 (0-10)	4	0 (0-4)	0
HRA R2	Jun 06	L	4 (0-10)	2	0 (0-4)	0
HYA R1	Dec 06	L	5 (0-11)	2	0 (0-4)	0
HYA R1	Jun 07	L	4 (0-10)	2	0 (0-3)	0
HRA R1	Sep 07	DU	3 (0-10)	3	0 (0-4)	1
HYA R2	May 08	DQ	3 (0-11)	2	0 (0-4)	1

* L = Linear; DL = Delayed Linear; DU = Delayed Uniform; D = Delayed Quadratic.

For the HPB/HNB reactors, much fewer axial cracks have been observed than at HYA/HRA, but there is a much larger number of circumferential cracks. Applying the statistical models to this class of cracks at these stations gives similarly satisfactory comparison between model predictions and observations. Currently, a Quadratic Model also provides the best fit to the available data, with the maximum likelihood model parameter values suggesting that risks from bore cracking have already ceased at these stations.

The statistical modelling also indicates that a significant fraction of circumferential cracking occurred more than ten years ago - this could be related to a 'weak' sub-population of bricks.

To date no keyway root cracks have been observed, but statistical methods have been developed for the prediction of these types of cracks. Thus, because of the variability of graphite properties over the core, it is expected that there will be a broad spread in the timing of these cracks if/when they occur.

PHYSICAL UNDERSTANDING

When graphite is irradiated in a reactor it initially shrinks, but later starts to expand as it passes through shrinkage turnaround. Material close to the fuel brick bore experiences a higher dose than material at the outside of the brick, so that early in life the bore shrinks at a faster rate than the periphery generating tensile stresses at the bore and compressive ones at the periphery. As the graphite at the bore approaches shrinkage turnaround, it shrinks at a slower rate than the graphite at the periphery generating compressive stresses at the bore and tensile ones at the periphery.

Even for the bore with its plain surface, the distribution of stresses at any time is a complex mixture of hoop and axial stress components that depends on the arrangements and sizes of their peripheral keyways (Figure 2). A further complication is that thermally-induced differential strains are generated during cold shut downs, and the magnitude and sense of these depend on the fast neutron irradiation reductions in Coefficient of Thermal Expansion (CTE). Again, because of the higher dose at the bore than the periphery, the graphite has a lower CTE at the bore than that at the periphery and this contributes to stress reversal earlier when shut down than when at power.

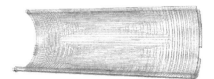 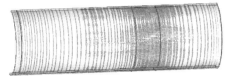

FIGURE 2: Bore Principal Stress Orientations Prior to Stress Reversal for HPB/HNB (left) and HRA/HYA (right) Fuel Bricks.

Graphite is about three times stronger in compression than in tension and so, given that the internal shrinkage and thermal stresses must be balanced, it is expected that cracks would predominantly initiate in regions of tensile stress. In AGR fuel bricks, these regions are at, or close to, the bore before stress reversal and at the periphery after stress reversal. Because the radial keyways only concentrate the hoop stresses, post stress reversal cracks would be most likely to initiate at the keyway roots and be axial.

Early calculations of the stresses in the AGR fuel bricks showed that the peak tensile stresses at the bore are about a fifth of the fracture strength of graphite so that, even allowing for variability and uncertainty in material behaviour, no cracking at the bore would be expected. However, these calculations were made to predict post stress reversal tensile stresses at the keyway roots and were simplified either by using two-dimensional approximations or by not including internal features such as the rings of vertical diffusion (methane) holes. In addition, the calculations assumed that the fuel remained at the centre of the channel.

Shortly after the first large, bore distorting crack was observed, an assessment of the stress concentration effects of the diffusion holes of the inner ring showed that the stress to strength ratio in the hoop direction at the inner surface of the methane holes would be about twice that at the adjacent position on the bore. This still leaves the peak stresses about 40% of the fracture strength of graphite and would, even so, only promote the initiation of predominantly axial cracks.

A closer inspection of the predicted bore stress distribution showed that there are significant regions where the hoop and axial stress components have comparable magnitudes. There is a significant body of experimental evidence showing that the effective strength of graphite under such biaxial stress conditions is lower than its strength under uniaxial stress conditions by around 20-30%. This is still not sufficient to explain the occurrence of bore cracks, but it would mean that there would be no particular preference for the initiation of axial, hoop or oblique cracks at the bore in these regions. Furthermore, the highest stresses would be expected to occur first and most frequently in the highest rated layers, *i.e.* in bricks in layers symmetrically above and below the centre of the moderator. This, too, is at odds with the observed bias of cracked bricks in the upper moderator layers and suggests that other factors contribute to the cracking process.

Increased TV coverage of the bore following the discovery of the first cracks at Hartlepool revealed light marks at the ends of fuel bricks in the upper moderator layers. These witness marks are a series of short horizontal stripes with the same separation as the ribs on the outer surfaces of the graphite fuel sleeves, suggesting that the fuel was in close proximity to the bore at these positions. Their positions in the core are also consistent with the fuel stringers having become bowed between the fuel supports at the bottom and the lateral restraints at the top provided by the upper stabilising brushes at Hinkley Point B and Hunterston B and the graphite seals at Hartlepool and Heysham 1.

The small movement of a fuel element within the channel has a negligible effect on the energy deposition in the brick but it disturbs the re-entrant (downward) gas flow in the annulus between the fuel element and the fuel brick. The mass flow past the bore on the side of closest approach of eccentric fuel is reduced, leading to less efficient heat convection and higher local brick temperatures with the reverse effects on the other side of the channel. These asymmetric temperatures can increase peak bore stresses by between 50 and 100% depending on the severity of the eccentricity changes.

The underlying cause of fuel eccentricity is a dose gradient across the diameters of the graphite sleeves in a stringer, leading to differential axial shrinkage and causing the stringer to bow away from the direction of highest dose. These conditions exist permanently in peripheral fuel channels where the fuel stringers bow out in the core radial direction, but only intermittently in non-edge channels where the dose gradients vary frequently with fuel power variations as part of normal fuel management. Another source of fuel eccentricity changes in non-edge channels occurs after radial fuel shuffling when old fuel is replaced by partly used fuel from peripheral channels, as these stringers are invariably bowed.

Fuel stringer eccentricity, whether it occurs systematically as in edge channel or randomly as in non-edge channels, is a whole channel phenomenon and so affects bricks in all relevant layers equally and at the same times. While multiple cracked bricks have been observed in the same channel, there have been few channels with pairs of adjoining cracked bricks and none with triples. It is also worth noting that not all channels with cracked bricks have witness marks and a significant proportion of channels with witness marks have no cracked bricks.

By noting, for bricks subjected to similar operating conditions, the varying amounts of crack pull-ins from brick to brick and direct observations of cracking times between inspections, it may be inferred that bore cracking has occurred over an extended period since the start of operations with the highest rate of cracking in the peak rated brick layers having occurred from the early to mid nineties until around 2001 at Hinkley Point B and Hunterston B, and from the late nineties until around the present day at Hartlepool and Heysham 1. Both periods coincide with when the bore tensile stresses at power are predicted to reach a broad flat peak and the graphite strength at the bore is reducing as a result of radiolytic oxidation. Furthermore, recent observations of cracks, which can definitely be identified as having initiated recently, have been in lower rated bricks where the bore stresses peak correspondingly later and the rate (with respect to time) of strength reduction by weight loss is lower.

The relatively low level of cracking of around 10% of the population at risk suggests that the high tail of the stress distribution and the lower tail of the strength distribution just overlap during this period. The evidence from material property measurements, including strength tests, suggests that variability in graphite is Gaussian with standard deviations, σ, in the range 5 - 15% of their mean values, μ. The negligible likelihood of peak brick stresses being greater than $\mu+2\frac{1}{2}\sigma$ and graphite strength being lower than $\mu-2\frac{1}{2}\sigma$ for bricks of similar manufacture and operating conditions suggests that either the mean values of strength and peak stress must be close, or their distributions have significant, non-Gaussian, outlier populations. The possibility of significant variability in stress between bricks arising from changes in fuel stringer eccentricity has already been discussed, but there may be other sources of outlier stress and strength arising from the manufacturing process including poor quality of graphitisation and machining, mishandling and pre-existing defects.

Support for the existence of outlier strength distributions comes from graphite strength tests using a large number of samples where weak outlier populations of around 3-5% were observed, giving rise to the possibility of a bimodal Gaussian distribution with the main distribution termed the background mode and the subsidiary mode termed the disparate mode (Kennedy and Eatherly, 1986). Additional evidence for a weak outlier population has been found from the failure rates for graphite fuel sleeve feedstock during proof pressure testing immediately following manufacture and prior to machining. Around 4% of the total failed at loads significantly lower than $\mu-2\sigma$ with a failure distribution that is non-Gaussian and which would be better represented by a simple power-law. Recent experimental results and theoretical studies suggest that each component of the overall distribution describes the characteristics of a different failure mode:

- Most fractures are caused by the creation of a defect of critical size and shape by a combination of the growth of a sub-critical defect and coalescence of sub-critical defects when the material is subject to high stress or strain (Li and Marrow, 2008). These background mode failures are caused by initiating defects grown from micro-structural features and defects that are characteristic of the material and their distribution is Gaussian.
- A small proportion of fractures are caused by the presence of a pre-existing defect significantly larger than the critical defects associated with the background mode fractures so causing fast fractures to occur at loads much lower than those associated with background mode failures. This disparate mode depends on the probability of occurrence of large isolated defects in the highest stressed regions of the components and has a simple power law distribution.

Two corollaries follow from the possible presence of a disparate fracture mode:

1. Where the gap between peak stress and strength is predicted to close gradually and the strength distribution is a power law, cracking will occur over the whole period when the stresses are tensile. The highest cracking will occur during the period immediately before the time of the smallest gap between peak stress and strength and will then cease as the gap increases prior to stress reversal.

2. Each high stress region in a complex component will have its own disparate failure mode and so each will contribute to the disparate mode for the whole component with the result that there will be a mix of crack initiation positions and morphologies. By comparison, background fractures will initiate at the positions with the highest stress and the crack morphologies will be dominated by the prevailing stress patterns.

Both are consistent with the observational evidence, although other processes could contribute to the cracking.

CONCLUSIONS

Statistical models of bore cracking in AGR reactors are able to make consistently good predictions of the number and type of bore cracks that are seen at outages. The latest fitting of the statistical model parameters to the available data suggest that the peak risk from bore cracking has passed at the lead stations and that these risks may even have ceased completely. This is broadly consistent with understanding of the processes involved obtained from mechanistic modelling and physical reasoning, giving added confidence that the evolution of the cores is satisfactorily understood.

To date no keyway root cracks have been observed at any of the AGRs, but if/when such cracks are observed, it is anticipated that it will be possible to employ a similar combination of statistical modelling and physical understanding to demonstrate continued reactor safety as the reactors age.

Acknowledgements
The authors would like to thank British Energy Generation Ltd for permission to present this paper.

References
Davison, A. C. (2003). Statistical models. Cambridge University Press.
Kennedy, C. R. and Eatherly, W. P. (1986). The statistical characterization of tensile strengths for a nuclear-type core graphite. *IAEA Specialists Meeting on Graphite Component Structural Design*, JAERI.
Li, H. and Marrow, T. J. (2008). In-situ observation of crack nucleation in nuclear graphite by digital image correlation. Paper PVP2008-61136, Proceedings of PVP2008, *2008 ASME Pressure Vessels and Piping Division Conference*, July 27-31, 2008, Chicago.
Maul, P. R. and Robinson, P. C. (2006). Cracking Down. *Nuclear Engineering International*, 51, [621], 44-48.
Maul, P. R. and Robinson, P. C. (2007). A strategy for monitoring bore cracking in AGR graphite cores. In:- *Management of Ageing Processes in Graphite Reactor Cores* (Ed. Gareth B. Neighbour), 248-255, RSC Publishing, ISBN 978-0-85404-345-3.

Development of a Dynamic Young's Modulus Test Setup for Irradiated and Oxidised Graphite

Onne Wouters[6]
Nuclear Research & consultancy Group,
P.O. Box 25, 1755 ZG Petten, The Netherlands
Email: wouters@nrg.eu

Abstract

As part of the extensive characterisation programme of the graphite specimens in the Blackstone British Energy materials test reactor project, the dynamic Young's modulus is determined by means of a measurement of the sonic velocity in the graphite. Since this technique makes it possible to determine a mechanical property in a non-destructive way on small specimens of varying geometries it is a popular test in irradiation programmes. In the Blackstone project, application of this fairly standard technique is complicated by the small size of the specimens, their high degree of oxidation and hence porosity and the fragile nature of the material. This presentation will address the various difficulties that have been encountered in setting up the technique. Extra attention will be given to the analysis of the acoustic signal that has traversed the graphite.

Keywords

Dynamic Young's modulus, Materials test reactor, AGR graphite

INTRODUCTION

The Nuclear Research and consultancy Group

The Nuclear Research and consultancy Group (NRG), based in Petten at the North Sea coast of The Netherlands is the Dutch centre of excellence in nuclear technology. With its staff of around 350, NRG develops knowledge, products and processes for safe applications of nuclear technology in energy, environment and health.

The High Flux Reactor

At the centre of NRG's research activities stands the High Flux Reactor (HFR). The HFR is a tank-in-pool multipurpose test reactor with a thermal power of 45 MW. Its core arrangement of 9 x 9 positions contains 33 fuel elements, 6 control rods, 22 beryllium reflector elements and 20 in-core experimental positions. Peak flux levels are $3.0 \cdot 10^{18}$ m^{-2}s^{-1} for thermal and $2.1 \cdot 10^{18}$ m^{-2}s^{-1} for fast neutrons respectively. Nuclear heating for in-core positions ranges from 1 to 12 W/g.

The Hot-Cell Laboratories and Jaap Goedkoop Laboratory

Dismantling of irradiation rigs and post-irradiation examinations are typically performed in NRG's Hot-Cell Laboratories (HCL). The HCL comprise five concrete and twelve lead shielded cells. Furthermore four glove boxes are present for less radioactive characterisation. All lead shielded cells and the glove-boxes are dedicated to materials testing. Tests include amongst others tensile, flexural and compressive strength, fatigue, creep, impact and fracture

---

[6] Author to whom any correspondence should be addressed.

toughness, (re-) welding, thermal diffusivity, dilatometry, electrical resistivity, density and dynamic Young's modulus.

In 2007 NRG opened the Jaap Goedkoop Laboratory, named after a former scientific director of the reactor centre, dedicated to research activities on low radio-active specimens. Inside another four glove boxes are fully equipped to perform state-of-the-art graphite research.

The Blackstone MTR Project
In cooperation with NRG, British Energy has launched an extensive materials test reactor (MTR) programme consisting of several irradiations of typical core material in the HFR up to neutron doses beyond that of the leading stations. The aim of this so-called Blackstone project is to extend - in preparation of upcoming safety cases - the knowledge of graphite properties to beyond what is currently in the database. The oxidation and temperature conditions in the HFR will resemble those in the AGR. To achieve this, the capsules will be continuously purged with a mixture of CO_2, CO and other gasses.

As in every MTR experiment the specimens are small. Various geometries are used, the smallest being a cylinder of 5 mm diameter and 6 mm height. The samples are so small for a number of reasons, the most important being the limited volume in an irradiation capsule combined with the wish for large numbers of data. Another reason to choose small specimens is the trepanning origin of most of the material which causes the samples to have an inherent gradient of weight loss and neutron dose.

NRG has a lot of experience in testing of nuclear graphite. None of these materials however have suffered from radiolytic oxidation. Questions can be raised whether the standard test methods are suitable for graphite, which has a considerable weight loss. To validate the measurement techniques at NRG and to acquire enough experience handling the low-density graphite, a thermal oxidation project has been executed, in which the radiolytic oxidation is simulated by thermal oxidation in a furnace. The results showed that most of NRG's techniques are applicable to highly oxidised graphite, but for some techniques modifications to the set-ups had to be made. This is especially true for the DYM set-up as will be explained later. This paper will focus on the work done to improve the old DYM setup and to make it suitable for the specimens in the Blackstone project. Results will be shown that prove that the technique can be applied successfully and repeatable on small specimens, given that the obtained data is processed in a well-defined way.

DYNAMIC YOUNG'S MODULUS BY TIME-OF-FLIGHT

Theory
It is not the aim of this paper to go into the physics behind the relationship between the sonic velocity in a material and its elastic constants (nor is it the expertise of the authors). For this paper it suffices to say that according to ASTM standard C769 the Young's modulus of a material can be determined by measuring the sonic velocity and using:

$$E = \rho v^2 \frac{(1+\sigma)(1-2\sigma)}{(1-\sigma)} \qquad [1]$$

with ρ the density of the material, v the sonic velocity in the material and σ is Poisson's ratio. The Poisson correction in formula 1 amount to 0.9 for $\sigma = 0.2$. Poisson's ratio however is not measured or known for irradiated oxidised graphite and 0.2 is used by lack of better information.

Application of this technique can be very useful in MTR programmes for various reasons. First, it gives ('pseudo') mechanical data without the necessity of performing a destructive test. Second, it can be applied on small specimens of various shapes. This makes it possible to perform DYM measurements on specimens that are designed to be used for a completely different test. Especially for post- and pre-irradiation comparisons the technique has proven very useful.

Experimental Set-Up

Using Equation [1], determining the DYM is just a matter of accurately determining the velocity of an acoustic signal in the graphite. This is achieved by placing the specimen in between two acoustical transducers. The transducers used at NRG are made from a piezo-composite and especially designed to be used on materials with high sound attenuation due to sound scattering or high surface roughness. This makes them ideally suited for use on the highly porous oxidised graphite. To achieve a good acoustic coupling between the transducer and the graphite, the probe is coated with a 1 mm stiff rubber layer. The resonance frequency at which these transducers operate varies between probes and can be tailored to the material under investigation. For non-oxidised graphite, NRG uses 5 MHz probes. The higher the frequency is, the more accurate the determination of the time-of-flight will be due to the smaller wavelength. High frequencies, however, get attenuated much more easily than low frequency signals. When the 5 MHz transducers were used on thermally oxidised graphite, hardly any signal was transmitted and no decent measurement could be made. Therefore the choice has been made to use 1 MHz transducers for the Blackstone project. This required an additional improvement on the acquisition system and analysis procedure.

One of the probes is linked to a pulse generator, which sends a very short (~100 ns) high voltage (~300V) pulse to one of the transducers, which consequently emits a small wave package through the graphite specimen. This wave package is received by the other transducer and converted to a voltage signal which is displayed on a high sampling rate digital oscilloscope. The high sampling rate is a pre-requisite since the time-of-flight can be as small as 1-2 µs. Although the signal from the pulse generator is also used to trigger the oscilloscope, and therefore the time between pulse generation and recording is exactly known, this is not equal to the traverse time in the graphite, for the largest part this is because of the time necessary for the signals to pass both rubbers. To overcome this difficulty, the signal is compared with a signal obtained in a situation without a specimen present, *i.e.* with both transducers in contact with each other. Typical wave packages for these reference measurements and measurements obtained on graphites are shown in Figure 1.

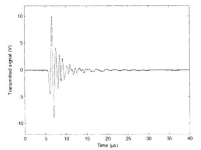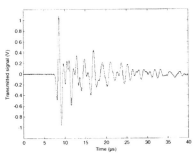

FIGURE 1: Signals shapes of left: reference signal (no graphite) and right: piece of irradiated and oxidized graphite.

Data Analysis

To analyse the data signals, they are first filtered. The ability of the oscilloscope to average signals combined with the continuously emitting pulse generator already significantly increases the signal-to-noise ratio. Therefore, an easy filtering consisting of the removal of higher Fourier components suffices.

The traverse time of the signal in the graphite is determined by comparing the transmitted signal with the reference signal. There are several ways in which this can be done:

1. Determination of the time delay between the signal onsets for both measurements.
2. Determination of the time delay between the first peaks in both signals.
3. Determination of the time delay between a series of peaks in both signals. The traverse time is then calculated as the mean value of the individual delay times.

The first method is as prescribed by the standard, but is hard to put in practice, since it requires a robust definition of the onset time of the wave packet. Looking closely at the onset region, it can be observed that the voltage starts changing slowly. Especially when the signal strength is weak and therefore the noise level higher, it is almost impossible to determine the onset of the signal. It might be attempted to fit the signal to a mathematical function and determine the point where it intercepts the axis, but this would lack physical support.

Especially on the filtered functions, the position of the peaks can be determined with great accuracy. Still there is a problem, since it is found that the frequency of the transmitted signal is not the same for the reference measurement and the measurement on the graphite sample. In Figure 2, part of a reference signal and part of a measurement signal on a piece of graphite are overlaid and scaled such that the intensity of the first peaks is equal.

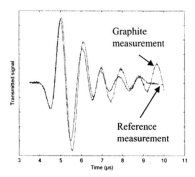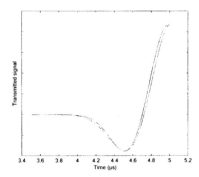

FIGURE 2: Comparison between the signal that has traversed a graphite specimen and the reference signal showing the change in frequency. Left: raw data of the reference measurement and the graphite measurement. The latter has been scaled and shifted in time so the first peaks overlap. Right: detailed view showing the first peak of the filtered signals.

From this picture it is clear that the signal that has traversed the graphite has a slightly lower frequency than the reference signal. This phenomenon is observed in all measurements and is more prominent in more oxidised graphite as is shown in Figure 3. This figure shows a reference signal and a measurement signal on a piece of PG-25 filter graphite. This material has a high porosity (density of ~1.2 g/cm^3). The traces are depicted at the same scale and the difference in frequency is large.

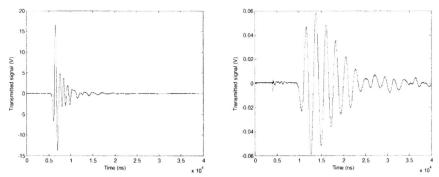

FIGURE 3: Reference signal (left) and measured signal (right) of a piece of PG-25 filter graphite. This highly porous graphite fully changes the frequency content of the transmitted signal.

The most likely explanation is that the more porous the material becomes, the more the signal is attenuated. Since the higher frequencies are more easily attenuated, the transmitted signal will slowly lose its higher frequency components. For less porous graphites only part of the dominant 1 MHz component will be damped and the spectrum of the resulting signal will be shifted only slightly. When the material is so porous that the entire 1 MHz component is damped, the transmitted signal, although low in amplitude, will be of a totally different frequency.

Clearly a comparison of the peak positions leads to an overestimation of the traverse time and therefore to an underestimation of the sonic velocity and the dynamic Young's modulus. The more peaks are taken into account, the larger this error will be. The error caused by this delay is smallest closest to the signal onset. Therefore, a compromise has to be found between the frequency change and the difficulties relating to the onset determination. NRG decided to determine the delay time at the point at which the first peak has reached 5% of its maximum value. In this way the advantages of the accurate peak determination are still used and the traverse time is determined close enough to the real signal onset not to worry about the frequency change but far enough away to determine a well defined value. All the raw data files will be stored, no information will be lost during the analysis. If insights progress, the measurements can be analysed at a later time.

VALIDATION

A lot of effort has gone into validation of this method against the ASTM standard. It has been investigated how repeatable the measurements is, what the influence is of different signal frequencies, different sample geometries (both diameter and length), different loads on the sample and how well the method is able to reproduce supplier values of a standard material. The most important results of this validation process will be discussed in the remainder of this paper.

Frequency Dependence
To check the influence of the signal frequency on the final time-of-flight, eleven specimens are measured with both 5 MHz and 1 MHz probes. As explained above, applying a 5 MHz

signal only works for non-porous graphite. The resulting sonic velocities are given in Table 1.

TABLE 1: Comparison between sonic velocities obtained using a 5 MHz probe
and a 1 MHz probe for a variety of non-irradiated and non-oxidised graphites.

Specimen	1MHz speed (m/s)	5MHz speed (m/s)
1	2606	2671
2	2718	2715
3	2735	2816
4	2678	2750
5	2484	2490
6	2495	2547
7	2346	2357
8	2455	2495
9	2252	2299
10	2845	2853
11	2560	2594

The average difference is smaller than 1.5 %. This collection of specimens contains graphites based on both pitch and petroleum coke, extruded and iso-moulded graphites and the grain sizes range from several tens of micrometers to close to a millimetre. This also means that the sometimes imposed criterion that the grain size of the graphite has to be small in comparison with the wavelength of the signal is not universally valid. In at least six of the above specimens the grain size is larger than the wavelength at 5 MHz.

Geometry Dependence
To see if the height of the specimens is of influence on the results, a special 'staircase' piece of stainless steel is applied. This material has been measured for sonic velocity independently and is used as a standard material. The staircase shape allows measurements on the same material but with different heights with increments of 1 mm. The results of a measurement on all steps of this specimen is shown in Figure 4.

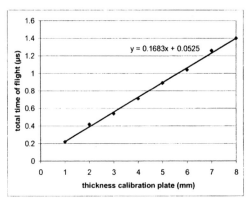

$y = 0.1683x + 0.0525$

FIGURE 4: Total time-of-flight for the different steps of the stainless steel staircase specimen. No deviations from the straight line are observed for smaller or larger measuring heights.

The measurements can nicely be fitted with a straight line, indicating no effect of sample length. The slope of the line gives the sonic velocity which matches the value given by the manufacturer. There is one issue showing up in these measurements: the line should pass the origin, but instead has a slight offset of 0.05 μs. This can be caused by a different impression on the rubbers between reference and steel measurements or by a misalignment of the two probes, which would lead to a small underestimation of the actual path travelled by the acoustic signal.

To assess the influence of the diameter of the specimen, two Gilsocarbon discs of 19 mm diameter and heights of 6 and 9 mm were measured. After the measurements were completed, five 5 mm cylinders were machined out of each disc. Measurements performed on the small cylinders were then compared to the full disk results. The results are summarised in Table 2.

TABLE 2: Results from size effect study. Two 19 mm diameter trepan slices are measured and consequently machined into five 5 mm diameter specimens. The measured velocity does not change.

Original	v (m/s)	Original	v (m/s)
Ø19 x 6 mm disc	2709	Ø19 x 9 mm disc	2718
Ø5 x 6 #1	2694	Ø5 x 9 #1	2711
Ø5 x 6 #2	2702	Ø5 x 9 #2	2708
Ø5 x 6 #3	2714	Ø5 x 9 #3	2706
Ø5 x 6 #4	2717	Ø5 x 9 #4	2701
Ø5 x 6 #5	2685	Ø5 x 9 #5	2690
average small samples	2702	average small samples	2703

A few things can be concluded from these results: First, there is no dependency on the cross sectional area of the specimens. The 19 mm disc is covering the entire area of the probe and the smaller specimens have the smallest load bearing area of all geometries that will be tested within the Blackstone project. Second, there appears to be no dependency on the length of the specimens. Since these two disks are of the same source material, the velocity of sound can be expected to be the same. Third, since the spread in the five (or even ten) individual results on the smaller specimens is very small, the influence of the local microstructure is not large. This is good news, since it means the small sample size as compared to the grain size is not affecting the accuracy of the measurement significantly.

CONCLUSIONS

The most important conclusion that can be drawn from this analysis of the DYM technique is that it is more independent of variations in experimental conditions than often suggested. The current ASTM standard sets limitations on frequency, graphite grain size, specimen length and specimen diameter without experimental support. The measurements performed in the frame of the Blackstone project show that these limitations are not always justified and this should be reflected in the next revision of the standard.

Especially because of its consistency when applied to small specimens, the DYM technique is very suited to be part of a characterisation programme in an MTR experiment. It is however very relevant that proper and well-defined data analysis is performed. Ideally all data should be recorded and stored for further analysis.

Stress Measurement in Porous Graphite at Different Length Scales

Soheil Nakhodchi[a], Peter E. J. Flewitt[b], David J. Smith[a]

[a]Solid Mechanics Research Group, Department of Mechanical Engineering,
University of Bristol, Bristol BS8 1TR
[b]H.H. Wills Physics Laboratory, University of Bristol, Tyndall Avenue, Bristol, BS8 1TL
Email: s.nakhodchi@bristol.ac.uk

Abstract

It is important to be able to measure internal stresses in the moderator bricks used in the UK gas cooled reactor cores. Over the operating life, the graphite is subjected to radiolytic oxidation and as a consequence the porosity can alter from levels of about 20%. The associated low elastic modulus, brittleness and relatively high porosity make it difficult to measure stress within these bricks. Two techniques have been used, incremental centre hole drilling (ICHD) and a deep hole drilling (DHD). Using known applied loads, virgin Pile Grade "A" and Gilsocarbon polygranular graphite are tested respectively. PG25 graphite with 48% porosity has been selected to represent a high level of porosity typical of that in irradiated graphite. The first technique provides a measure of subsurface stresses to a depth of 1 mm and the latter is performed to obtain through thickness stress measurements. This demonstrates that standard strain gauge ICHD can estimate near to surface stresses in virgin Gilsocarbon graphite; however, in the porous graphite stresses were underestimated. The application of the DHD method illustrated that the technique has a potential to perform through thickness measurements on graphite core reactor up to at least 48% percentage porosity.

Keywords

Deep hole drilling, Incremental centre hole drilling, Polygranular graphite

INTRODUCTION

The reactor cores of gas cooled reactors which are used in the UK are made from assemblies of bricks of polygranular graphite. Exposure of these reactor cores to the service environment of fast neutrons and CO_2 coolant gas results in irradiation hardening and radiolytic oxidation of the graphite during the service life. These effects lead to progressive changes in the physical and mechanical properties of the bricks. The corresponding overall degradation of the graphite has to be evaluated to provide confidence in the continued safe operation of the plant.

A knowledge of applied and residual stresses is essential to assess integrity of engineering components. Applied stresses can be estimated from numerical methods, however, finding a reliable method for the determination of residual stresses continues to be a challenge for engineers and scientists. Measurement is one of the suggested solutions. A wide variety of residual stress measurement techniques at different length scales have been developed and applied for metallic alloy components (Lu, 1996). Although polygranular graphite is widely used in the reactor cores, there is relatively limited experimental work reported for *measurement* of residual stress. In this paper, residual stress measurements applied to graphite using incremental centre hole drilling (ICHD) and deep hole drilling (DHD) are described. The ICHD method is a semi destructive technique for near to surface residual stress measurement. This technique is a further development of Centre Hole Drilling (CHD) method to measure non-uniform stress profiles. In the ICHD technique the calibration coefficient has to be obtained prior to measurement, using either experimental or finite element method (Rendler *et al.*, 1966, Schajer, 1981). Among different data analysis procedures (Schajer, 1998) the integral method (Zuccarello, 1999) is the most common method for analysing the data obtained from the experiment. Recent developments of the ICHD method can be found in (Stefanescu *et al.*, 2006). DHD is a also a semi-invasive method for through thickness residual stress measurement. Initial studies by Zhdanov and Gonchar (Zhdanov *et al.*, 1978) were continued by Smith and co-workers to improve the technique and the method has been used to measure applied (George *et al.*, 2002) and residual stresses in metallic (George *et al.*, 2005, Smith *et al.*, 2000) and composite (Bateman *et al.*, 2005) materials successfully. Recently, the versatility of the technique for measuring residual stress distributions in components with difficult access to the measurement location and in-situ measurement for full-scale components has been reported by (Kingston *et al.*, 2006).

In this paper, a series of stress measurement studies on polygranular graphite at different length scales are described, the ICHD and DHD technique. The aim of the study was to explore applicability of these techniques in graphite with the current standard and procedure used in isotropic metallic materials.

SUMMARY OF THE TECHNIQUES

The analysis used in the ICHD and DHD techniques is summarized here. More details can be found in (Zuccarello, 1999) and (Bonner, 1996), respectively.

Incremental Centre Hole Drilling

The strain gauge ICHD technique involves drilling a small hole, about 2 mm diameter, in a series of steps with gradually increasing the hole depth. Relaxed strains are measured after each increment step, using a rosette strain gauge. The special configuration of rosette strain gauge pattern designed for ICHD typically includes three strain gauges oriented in the specific directions, $0°$, $45°$, $90°$ (Rendler *et al.*, 1966). This allows a strain reading of $\varepsilon_1, \varepsilon_2$ and ε_3 after each drilling step. In the analytical approach strains relaxed after each drilling step depend on stresses existing at all hole depths up to that point. In integral method (Schajer, 1998), strain relaxation due to drilling along a generic x direction can be expressed in the form of (Zuccarello, 1999).

$$\varepsilon_{xn} = \frac{1}{2E} \sum_{i=1}^{n} \left[(1+v)a_{ni}(\sigma_{xi} + \sigma_{yi}) + b_{ni}(\sigma_{xi} - \sigma_{yi}) \right] \qquad \text{n=1......N} \qquad [1]$$

where E is the Young modulus, v is the Poisson ratio and n is the depth increment, σ_{xi} and σ_{yi} are stresses equivalent to the uniform residual stress within the ith layer and a_{ni}, b_{ni} are the calibration coefficient provided by finite element analysis (Schajer, 1988). Defining the following parameters:

$$p_n = \frac{\varepsilon_1 + \varepsilon_2}{2}, \quad q_n = \frac{\varepsilon_1 - \varepsilon_2}{2}, \quad t_n = \frac{\varepsilon_1 - 2\varepsilon_2 + \varepsilon_3}{2} \qquad [2]$$

$$P_n = \frac{\sigma_x + \sigma_y}{2}, \quad Q_n = \frac{\sigma_x - \sigma_y}{2}, \quad T_n = \frac{\sigma_x - 2\tau_{xy} + \sigma_y}{2} \qquad [3]$$

where p_n, q_n and t_n are strains and P_n, Q_n and T_n are stresses calculated for each drilling step (Zuccarello, 1999) Equation [1] can be solved and rewritten in matrix form as:

$$\{P\} = \frac{E}{1+v}[A]^{-1}.\{p\} \qquad \{Q\} = E[B]^{-1}.\{q\} \qquad \{T\} = E[B]^{-1}.\{t\} \qquad [4]$$

where $\{P\}, \{Q\}$ and $\{T\}$ are stress vectors, $\{p\} \{q\}$ and $\{t\}$ are strain vectors. $[A]$, $[B]$ are calibration coefficient matrices. Finally principle stress and strain vectors, σ_{max} and σ_{min}, and their orientation, β, can be calculated as:

$$\sigma_{max}, \sigma_{min} = \{P\} \pm \sqrt{\{Q\}^2 + \{T\}^2} \qquad [5]$$

$$\beta = \frac{1}{2}\arctan\left(\frac{\{T\}}{\{Q\}}\right) \qquad [6]$$

Deep Hole Drilling
There are four basic steps in the DHD method. First, bushes are glued on to surfaces of the sample at the entrance and exit points of the reference hole. Second, a reference hole is drilled through the sample and the bushes. The hole diameter is then measured using an air probe system, with measurements being taken at many points along the length of the hole and at many angles. To release the stresses a core containing the reference hole is trepanned and after trepanning the hole diameter is re-measured.

Using the measurement of the hole distortion between stressed and unstressed states, permits the DHD technique to measure residual stress distributions in a component. The measured radial distortions of the hole, $\bar{u}$, are related to the residual stresses, $\bar{\sigma}$, during the trepanning using:

$$\bar{u} = -[M]\bar{\sigma} \qquad [7]$$

where the hole distortion and the applied stress vectors are given by

$$\bar{u} = [u_0(\theta_1, z_1), u_0(\theta_2, z_1), ..., u_0(\theta_n, z_1), u_{zz}]^T \qquad \text{and} \qquad \bar{\sigma} = [\sigma_{xx}, \sigma_{yy}, \sigma_{zz}]^T$$

where z is the depth through the thickness. The compliance matrix M is given by

$$[M] = \frac{1}{E}\begin{bmatrix} f[\theta_1, z_1] & g[\theta_1, z_1] & h[\theta_1, z_1] \\ f[\theta_2, z_1] & g[\theta_2, z_1] & h[\theta_2, z_1] \\ \cdot & \cdot & \cdot \\ \cdot & \cdot & \cdot \\ \cdot & \cdot & \cdot \\ f[\theta_n, z_1] & g[\theta_n, z_1] & h[\theta_n, z_1] \end{bmatrix}$$

Since the distortions are measured Equation [7] has to be inverted to obtain the stresses. Therefore,

$$\hat{\sigma} = [M]^*\bar{u} \qquad [8]$$

where $[M]^+ = (M^t M)^{-1} M^t$ is the pseudo-inverse of the matrix [M], and $\hat{\sigma}$ is the optimum stress vector that best fits the measured distortions. Finally, the residual stress distribution through the thickness is obtained by measuring the distortions of the reference hole using Equation [8] at each measurement position z.

MATERIALS AND SPECIMENS

Three grades of polygranular graphite were used in the experiments, Pile Grade "A" (PGA), Gilsocarbon and PG25 porous filter graphite. The PGA samples extracted from blocks of extruded materials used in an earlier programme by Holmes (Holmes, 2003) has anisotropic material properties. To obtained graphite with different elastic modulus, E, rectangular beam specimens were cut in directions parallel and perpendicular to brick axis as shown schematically in Figure 1a. The mean value of E in the parallel direction is about 11.3 GPa and is about 3.4 GPa in the perpendicular direction. A rectangular bar of 190 mm length with square section of 22 mm depth and 23 mm width was extracted parallel to the graphite brick axis. A further PGA graphite, rectangular bar of length 190 mm, depth 25 mm and width 35 mm, was extracted from the same blocks but perpendicular to brick axis. Gilsocarbon graphite has isotropic material properties and rectangular bars of length 190 mm, depth 25 mm and width 35 mm were cut from a solid cylinder provided by British Energy Ltd. The cutting direction is shown in Figure 1b. As fabricated, both the PGA and AGR have up to about 20% porosity. PG25 porous filter graphite, with 48% porosity procured from Morganite Electrical Carbon has a level of porosity that is encountered in irradiated graphite. Rectangular bars of 180mm length, 25mm width and 35mm depth were cut from this block. Application of the DHD technique was examined using PGA and PG25 porous graphite, whereas, AGR and PG25 were employed for the ICHD experiments.

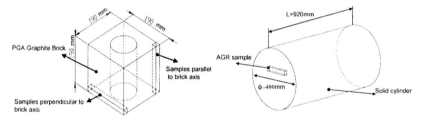

FIGURE 1: The orientation of sample extracted from a) a PGA graphite brick b) a Gilsocarbon solid cylinder.

EXPERIMENTS

The experiments were conducted on the rectangular specimens, using a four point bend rig to introduce a known stress field, using an in-house designed and manufactured rig. As illustrated schematically in Figure 2, the region between the two innermost rollers of the jig is subjected to a constant bending moment. This provides a linear distribution of stresses through the depth, h, such that tensile stresses are at the top and compressive stresses are at the bottom surface of the beam. ICHD measurements were performed at the top and side bottom surfaces of the samples to investigated accuracy of the technique in measurement of the near surface tensile and compressive stresses. The DHD method was applied through the thickness of the samples. Figure 3 shows locations on the samples that where measurements were carried out. Different specimens were used in each ICHD test and strain gauges shown in Figure 3a were bonded to different specimens which are shown in one figure.

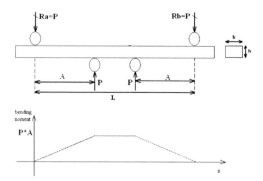

FIGURE 2: Four-point bend stress field.

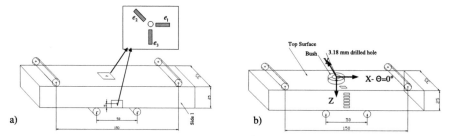

FIGURE 3: Location of the measurement in a) ICHD samples and b) DHD samples –all dimensions are in mm.

ICHD Measurements

For the ICHD measurements, a 062UL rosette strain gauge was bonded to the surface of the bend specimens at the locations shown in Figure 3a. M bond 200 was adhesive was used for attaching strain gauges to the Gilsocarbon graphite sample, however due to high level of porosity of PG25 a more viscous adhesive, AE10, had to be used for PG25 graphite. First, samples were loaded in a four points bend rig and using Hooke's law applied stresses were determined directly from the strain gauge readings. Then, the rosette strain gauges were zeroed to start the ICHD measurements. A 1.6 mm diameter drill was used with a conventional Measurement Group RS200 drilling rig to produce a hole at the geometric centre of the rosette in a series of steps. Additional uniaxial strain gauges were bonded to the top and bottom surfaces of the samples to control the load. ICHD measurements were undertaken on the Gilsocarbon samples and PG25 porous graphite specimens.

DHD Measurements

PGA and PG25 beam specimens were cut parallel and perpendicular to brick axis. Prior to DHD measurement, each specimen had reference bushes glued to the top and bottom surfaces, Figure 3b. Then a 3.18 mm diameter reference hole was machined into the centre of the beam with the axis parallel to the direction of loading. For each test specimen, two strain gauges were bonded on the top surface and two on the bottom to monitor the load. Instead of carrying out the trepanning process used in the conventional DHD method to relax the residual stresses the diameter of the through thickness reference hole was measured using an air probe when the beams were unloaded and loaded. Hole distortions were measured at intervals of 0.2 mm through the depth and at intervals of 20 degrees around the circumference.

RESULTS

Relaxed strains measured during ICHD test in a Gilsocarbon specimen are shown in Figure 4. The data points are experimentally measured strains which are shown along with their interpolating curves from which the strain components p, q and t were calculated. The stresses obtained using integral method explained above. In practice different order polynomials (3rd to 8th order) were fitted to the strain data and for each of them value of the stresses were calculated using Schajer's cumulative influence functions (Schajer, 1998). In all results presented, raw strain data were used with averaging stresses obtained from each of the strain fitted curves.

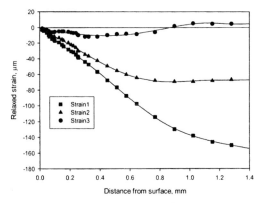

FIGURE 4: Relaxed strains in tensile region in Gilsocarbon graphite specimen, points are measured strain and lines is a sixth order polynomial fitted to these data.

Figure 5a shows measured tensile stresses in the Gilsocarbon graphite specimen compared with the applied stresses. Using strain gauge readings and elastic beam theory, the applied tensile stress was expected to vary from 5 MPa at the surface to 4.6 MPa within the 1 mm below the surface. Compressive stresses were also measured on the side of the specimens. Elastic beam theory predicted a constant value of stress through the specimens, Figure 3a. This is illustrated as continuous line in Figure 5b which is compared with the measurement points. Schajer (Schajer *et al.*, 1996) has shown that ICHD measurements are very sensitive to strain error as depth increases. Small strain errors at the surface cause large errors. An example of including an error of ±0.5με in strain reading and propagating it to the stresses is illustrated in Figure 6.

ICHD results for tests calculated using PG25 shown in Figure 7, where both tensile and compressive stresses were underestimated. Although the reason for this is not fully known, it is anticipated that effect of porosity are significant since in PG25 the volume of fraction was 48%. This value in Gilsocarbon was about 20%, *i.e.* with the same drilled hole diameter the amount of bulk material removed by drilling in PG25 is calculated to be 0.6 times that in the Gilsocarbon graphite. On the other words, equations used in standard ICHD technique for isotropic material appear to be valid for graphite with a fraction volume of up to about 20% and for a higher void volume fraction the calibration coefficients may need to be modified.

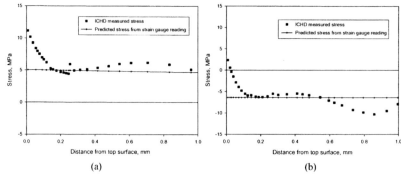

(a)   (b)

FIGURE 5: Surface stress measurements in gilsocarbon specimens a) tensile region b) compressive region.

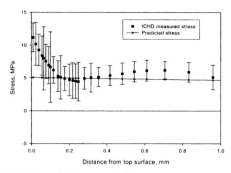

FIGURE 6: Errors in ICHD resulted from propagating ±0.5με error.

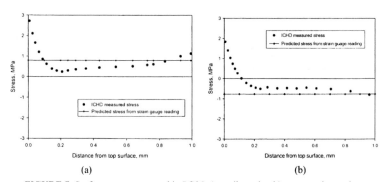

(a)   (b)

FIGURE 7: Surface stress measured in PG25 a) tensile region b) compressive region.

From the DHD measurements in the graphite specimens, a typical angular hole distortion at three various depths from the top surface is shown in Figure 8. Points are measurement values and curves represent fits to the experimental normalized hole distortion. Stresses were then determined based on Equation [8].

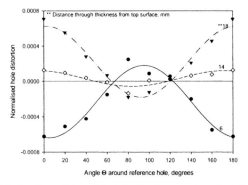

FIGURE 8: Angular variation of the hole distortion at various depth from the top surface of the PGA specimen cut parallel to brick axis.

In Figure 9a through thickness direct stresses, σ_{xx}, for PGA samples extracted parallel and perpendicular to brick axis are compared with surface strain gauge values. Both test specimens were subjected to almost the same level of strain. The different levels of stress illustrated in the Figure 9a are due to difference in the modulus of elasticity. E in the parallel direction was higher than in the perpendicular direction so that the stresses were higher in the parallel specimen. The hole distortion was measured in the PG25 porous graphite specimen but unlike the PGA specimens the presence of the porosity increased variability of the measured diameter. Indeed, hole with the diameter bigger than the actual drilled hole was measured. Averaging measured hole diameters over the 1 mm intervals through the depth reduced the variability in the measured hole distortion. Also irrespective of the generation of a dummy hole in the PG25 porous graphite, as the DHD technique is based on the *deformation* of the reference hole not the diameter of the actual hole, it can be seen that, Figure 9, the calculated stresses compared well with the surface strain gauge stresses.

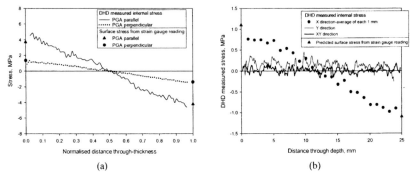

(a)   (b)

FIGURE 9: Through thickness measured stress a) PGA parallel and perpendicular b) PG25 porous graphite.

Further study is being conducted on application of the air-probe for estimation the pore diameter in graphite with different levels of porosity.

CONCLUDING REMARKS

Application of two methods, ICHD and DHD, for measuring stresses in graphite at the different length scales, up to 1mm below the surface and though the thickness, has been investigated this shown that using the standard strain gauge ICHD method, applied tensile and compressive surface stresses can be estimated in Gilsocarbon graphite. However, application of the technique to highly porous PG25 graphite tended to underestimate the applied stresses in both tensile and compressive regions. This can be related to the relatively high volume of fraction of porosity in the materials.

The DHD method when applied to the PGA graphite with different values of Young's modulus shows a good agreement with the stresses obtained from strain gauges. Using material with higher volume of fraction, PG25, 48% porosity, it is still practical, to measure applied stresses.

Acknowledgements
This work was carried out as part of the TSEC programme KNOO and as such we are grateful to the EPSRC for funding under grant EP/C549465/1.

References
Bateman, M. G., Miller, O. H., Palmer, T. J., Breen, C. E. P., Kingston, E. J., Smith, D. J. & Pavier, M. J. (2005). Measurement of residual stress in thick section composite laminates using the deep-hole method. *International Journal of Mechanical Sciences, 47*, 1718-1739.
Bonner, N. (1996). Measurement of residual stresses in thick steel welds. *Mechanical Engineering*. Bristol, Bristol.
George, D., Kingston, E. & Smith, D. (2002). Measurement of through-thickness stresses using small holes. *The Journal of Strain Analysis for Engineering Design, 37*, 125-139.
George, D. & Smith, D. J. (2005). Through thickness measurement of residual stresses in a stainless steel cylinder containing shallow and deep weld repairs. *International Journal of Pressure Vessels and Piping, 82*, 279-287.
Holmes, C. (2003). The mechanical behaviour and ultrasonic measurement of graphite joints. *Mechanical Engineering*. Bristol, University of Bristol.
Kingston, E. J., Stefanescu, D., Mahmoudi, A. H., Truman, C. E. & Smith, D. J. (2006). Novel applications of the deep-hole drilling technique for measuring through-thickness residual stress distributions. *Journal of ASTM International, 3*.
Lu, J. (1996). *Handbook of measurement of residual stress* Society for Experimental Mechanics.
Rendler, N. & Vigness, I. (1966). Hole-drilling strain-gage method of measuring residual stresses. *Experimental Mechanics, 6*, 577-586.
Schajer, G. S. (1981). Application of finite element calculations to residual stress measurements *ASME Journal of Material and Technology, 103*, 157-163.
Schajer, G. S. (1988). Measurement of non-uniform residual-stresses using the hole-drilling method .1. stress calculation procedures. *Journal of Engineering Materials and Technology-Transactions of the Asme, 110*, 338-343.
Schajer, G. S. (1998). Measurement of non-uniform residual stresses using the hole drilling method. part II: practical application of the Integral Method. *Trans. ASME, Journal of Engineering Materials and Technology, 46*.
Schajer, G. S. & Altus, E. (1996). Stress calculation error analysis for incremental hole-drilling residual stress measurements. *Journal of Engineering Materials and Technology, 118*, 120-126.
Smith, D., Bouchard, P. & George, D. (2000). Measurement and prediction of residual stresses in thick-section steel welds. *The Journal of Strain Analysis for Engineering Design, 35*, 287-305.
Stefanescu, D., Truman, C. E., Smith, D. J. & Whitehead, P. S. (2006). Improvement in residual strees measurement by the incremental hole drilling. *Experimental Mechanics, 46*, 417-427.
Zhdanov, I. M. & Gonchar, A. K. (1978). Determining the residual welding stresses at a depth in metals. *Automatic Weld, 21*, 22-24. .
Zuccarello, B. (1999). Optimal calculation steps for the evaluation of residual stress by the incremental hole-drilling method. *Experimental Mechanics, 39*, 117-124.

Mechanical and Thermal Property Measurements on AGR Core Graphite using Electronic Speckle Pattern Interferometry

Paul Ramsay[7] and John G. Gravenor

National Nuclear Laboratory, Sellafield, Cumbria, UK. CA20 1PF
Email: paul.ramsay@nnl.co.uk

Abstract

The strength and integrity of the graphite moderator in Advanced Gas-Cooled reactors is an important consideration for safe operation and plant life extension. Accordingly samples of the core graphite are routinely taken for monitoring of the physical properties, including the coefficient of thermal expansion and the elastic modulus. The experimental techniques currently employed have various limitations and so the possibility of using electronic speckle pattern interferometry (ESPI) has been explored for making measurements of displacement in small specimens under changes in load or temperature. ESPI is a high resolution technique which gives a detailed map of the full displacement field across the specimen. Using a commercial instrument, evaluation tests have been carried out on samples of unirradiated Gilsocarbon graphite. Measurements of the quasi-static elastic modulus using 3-point bend tests on miniature beams have been made, which have shown the need for careful test fixture design and beam uniformity to avoid the effects of friction and unsymmetrical loading. The high resolution of the technique allows the non-linear stress-strain characteristics of graphite to be studied and optimum methods of data analysis to be developed. The coefficient of thermal expansion of reactor graphite is currently measured in three different directions in three separate tests using dilatometers. Initial trials using the ESPI technique offer the prospect of simultaneous measurement in two directions and the determination of localised variations, but indicate that the furnace performance needs further assessment for this application.

Keywords

ESPI, Elastic modulus, Poisson's ratio, Coefficient of thermal expansion

INTRODUCTION

The physical properties of the moderator graphite in Advanced Gas-cooled Reactors (AGRs) are monitored by means of cylindrical samples extracted from the fuel channel bores. Among the properties measured are strength, dynamic elastic modulus and the coefficient of thermal expansion (CTE). The bored samples are 40 mm long and 19 mm diameter and with the requirement to measure properties related to the radial dose profile, miniature test specimens have to be used. Discs 6 mm thick and 19 mm diameter are used for dynamic modulus measurements using a time-of-flight ultrasonic method. For CTE measurement by dilatometry, the discs have flats machined on the sides. Finally, square-section beams 6 mm x 6 mm x 19 mm long are machined from the discs and tested in 3-point bending to determine the flexural strength. Various experimental problems have been encountered with both elastic modulus and CTE measurements. Dilatometry is time-consuming and sensitive to the pushrod-specimen contact characteristics, while transducer-specimen coupling effects, dispersion and data interpretation problems complicate the ultrasonic method.

---

[7] Author to whom any correspondence should be addressed.

Since elastic modulus and CTE determination essentially require the measurement of displacement under an applied load or temperature change, Electronic Speckle Pattern Interferometry (ESPI) has been investigated for these applications. Speckle arises when an optically rough surface is illuminated with coherent laser light, each speckle being the vector sum of the amplitude and phase of light scattered from each point of the surface. Combining the speckle with a reference beam gives a fringe pattern which is imaged by the camera. A similar image taken after deformation can be compared to derive the three dimensional displacement fields over the entire surface image. Thus, the technique is non-contacting and is capable of high displacement resolution, typically 100 nm, but down to 10 nm under favourable conditions.

DETERMINATION OF ELASTIC CONSTANTS AND CTE

For a linear elastic material the normal x-strain in terms of the stresses is given by:

$$\varepsilon_x = \frac{\sigma_x}{E} + \frac{v\sigma_y}{E} + \frac{v\sigma_z}{E} \qquad [1]$$

with similar relations for ε_y and ε_z

Since bending is a plane stress deformation, $\sigma_z = 0$ and eliminating v from the equations for ε_x and ε_y, E can be found from the strains measured on the side face of the beam if the stresses are known. At the bottom of the beam σ_y is zero at the free surface and so:

$$E = \frac{\sigma_x}{\varepsilon_x} \qquad [2]$$

The same procedure, this time eliminating E from the elasticity equations, gives the Poisson's ratio at the bottom surface as:

$$v = \frac{\varepsilon_y}{\varepsilon_x} \qquad [3]$$

The bending stresses from a concentrated load may be determined analytically (Timoshenko, 1970) and by finite element analysis. Future work will use improved modelling of the total non-linear stress field to enable the use of all the ESPI displacement data, and to calculate the elastic properties for all points of the sample. The CTE is found directly from displacement measurements made over a temperature interval.

IMPLEMENTATION

A number of evaluation trials have been carried out, using an ESPI system to monitor the displacements of the side of a graphite beam during a three point bend test. Different sized samples of unirradiated Gilsocarbon reactor grade graphite have been tested, with attached strain gauges for comparison purposes. These samples were subjected to stepped load increments with the ESPI camera taking speckle images at static load points. After testing, the ESPI software processes pairs of images to determine the dimensional changes relative to an arbitrary point in the field of view, for each of the pixels which make up the image. Each of these step changes are summed to provide the total displacements for each of the three x, y and z directions. These displacement arrays can then be used to determine the strain at any point by differentiating the displacement curves.

Vibration

Early tests showed that the ESPI technique is very sensitive to interference from vibration. Relative movement between the camera and sample of the order of the laser light's 780 nm wavelength causes the software to incorrectly determine the directional phase of the displacements. To minimise this effect the coupling of the camera mount to the testing machine must be made as rigid as possible and the machine itself be suitably isolated from sources of vibration.

Beam test fixture

Although the ESPI technique is capable of accurately measuring the beam strain distribution, the stresses are only known with similar accuracy if the test piece is subjected to symmetrical deformation. This requires that the beam is held firmly and accurately within the test fixture so that no twisting in the z-plane occurs. With the small sample dimensions and low strain to failure for graphite, this requirement can be difficult to achieve, especially if remote handing is required within a shielded facility for irradiated material.

RESULTS

10 x 10 x 50mm Beam

An example of ESPI measurements made in a test on a 10 x 10 x 50 mm beam using a 40 mm span is shown in Figure 1. The ESPI x-strain of 1.09 mm/m agreed with that obtained from the attached strain gauge, and the x-stress at the bottom surface was 11.2 MPa in response to the 195N load applied. The compressive y-strain was measured vertically under the load application roller; the upper part of the beam was locally distorted by the load contact, but the linear lower portion gave a y-strain of 0.23 mm/m. These values correspond to 10.3 GPa for Young's modulus and 0.21 for the Poisson's ratio at the intersection point.

6 x 6 x19mm Beam

This smaller beam size was evaluated to determine the suitability of ESPI on samples currently measured as part of the AGR monitoring programme. Figure 2 shows that the signal noise is larger, but the indications are that with the good resolution of the 13μm pixel size, ESPI is detecting the granular nature of the material.

The elastic properties measured for this sample were 12.3 GPa for Young's modulus (E) and 0.20 for the Poisson's ratio. The higher value for E is due to the unsuitable fixture used for these tests, in that the lower beam supports were fixed. Since the supports are unable to roll outwards as the beam bends, this causes frictional forces in the span direction which act to increase the apparent stiffness and reduce the bending stress; the effect will be proportional to the depth/span ratio of the beam. All 3-point bend testing carried out at the Windscale Laboratory is done with rollers for the supports, which can roll freely on a flat bed so minimising friction.

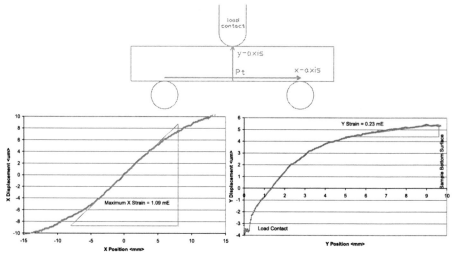

FIGURE 1: Typical response of 10x10x50mm Gilsocarbon graphite beam
following 195N (11.2 MPa) load increase.

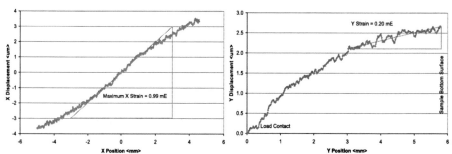

FIGURE 2: Typical response of 6x6x19mm Gilsocarbon graphite beam
following 126N (12.2 MPa) load increase.

Figure 3 shows the stress/strain curve for a 6 x 6 x 19 mm beam which has been cyclically loaded 4 times to approximately 50% of failure stress, over a period of 15 minutes. This sample was in the as-manufactured state (*i.e.* not previously stressed) and displays the permanent strain experienced by virgin graphite on the first loading. The trace also shows large hysteresis during the unloading part of the cycles.

Published literature describes the hysteresis between loading and unloading as 'elastic hysteresis' or 'damping' and is attributed to internal friction. In ceramics, this hysteresis is normally attributed to friction as microcrack boundaries experience shear forces. Following on from this explanation, the initial permanent strain observed would be created as new microcracks are formed and increase the volume of the sample.

The problem in using the microcrack hypothesis for graphite is that the higher density of cracks would also permanently lower the material modulus during subsequent load cycles. The present ESPI data shows that the Young's modulus at zero stress is constant between all of the load cycles. One possible scenario is that rather than the graphite experiencing actual cracking in which the bonding between grains is completely broken, the pure graphite crystals experience a shear deformation at the weak binding between the much stronger basal planes (Kelly 1981). This shear will be accommodated within the polycrystalline bulk material by a volume change, but the overall strength will not be changed since all the atomic bonds are unaffected.

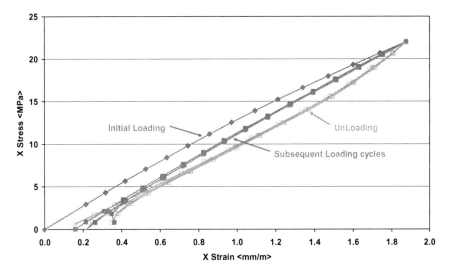

FIGURE 3: Stress/Strain response of 6x6x19 mm virgin Gilsocarbon graphite beam to cyclic loading.

Using the differences in stress and strain between each of the load increments allows the Young's modulus to be calculated at different stresses during the entire test, as shown in Figure 4. The first loading portion produces an E which decreases with stress at a greater rate than that of the subsequent cycles, but significantly, the modulus at low stress, <3MPa, is the same for both initial and later load cycles. The elastic hysteresis in the unloading portions produces an extreme variation in apparent modulus.

Further work will be required to determine if the change in Young's modulus with stress is real or a consequence of the friction between the beam and the test fixture supports.

Using the changes in the x and y strains between each load step allows the Poisson's ratio to be determined at different stresses. Figure 5 shows that overall there is very little variation. However, a single test to failure of a 10 x 10 x 50 mm beam does show that during the initial loading phase the Poisson's ratio decreases with stress as a consequence of the permanent strain accumulation; tensile x-strain accumulates faster than the compressive y-strain.

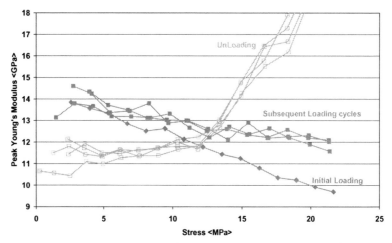

FIGURE 4: Young's modulus of 6x6x19mm virgin Gilsocarbon graphite beam as measured by ESPI during a cyclic loading test.

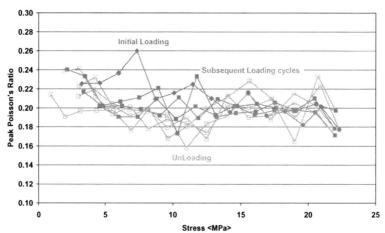

FIGURE 5: Poisson's ratio of 6x6x19mm virgin Gilsocarbon graphite beam as measured by ESPI during a cyclic loading test.

COEFFICIENT OF THERMAL EXPANSION

A trial has been carried out to examine the suitability of ESPI for the measurement of CTE in graphite. A sample of unirradiated graphite, prepared in the same manner as current CTE specimens, was examined using the ESPI manufacturer's standard hotplate furnace, normally used for electronic component thermal strain assessments. The sample was both ramped to 300°C and returned to room temperature at a constant rate of 5°C/min, while the ESPI instrument took images at 30 second intervals. Figure 6 shows the distribution of the x and y components of the relative displacements. The contours are parallel which indicates that the temperature distribution of the sample surface is uniform up to the edges although there will be a discrepancy between the hot plate and measured surface temperatures. The slight rotation of the contours indicates that the sample was rotating during the test caused by the furnace hotplate moving on its supports.

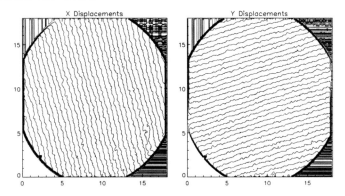

FIGURE 6: Relative displacement measurements, for a 275°C temperature change, of a 6mm thick disc of virgin Gilsocarbon Graphite. (1μm contours).

This test has again showed that ESPI is very sensitive to sample vibration, along with the Schlieren effects (optical inhomogeneities of the 'hot air' turbulence); this test had a noise level of 250nm, which will need to be improved for future work. This noise required excessive filtering and averaging to derive the instantaneous CTEs as shown in Figure 7. These data show that there is a large discrepancy between the heating and cooling phases which are caused by the hysteresis between the temperature of the hotplate and that of the top surface of the sample. Future work will require that the sample's top surface temperature is measured directly via an optical pyrometer so as to remove this effect. The overall values obtained by ESPI are also significantly higher than the values obtained from similar material, measured using a dilatometer. The source of this offset is currently unknown, but with the temperature dependent slope close to that of the 'cooling' curve it is expected that the discrepancy is due to pixel scaling problem.

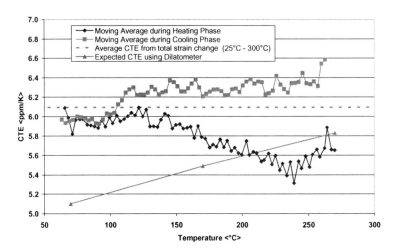

FIGURE 7: Measurement of Coefficient of Thermal Expansion of virgin Gilsocarbon Graphite using ESPI.

CONCLUSIONS

The use of ESPI as a non-contact measurement system for use in irradiated material testing shows great promise. The quality of the strain measurements is similar to that obtained with other techniques, but without the difficulties of sample preparation and transducer attachment and with the benefit of simultaneous measurement in more than one direction.

The values obtained by ESPI for the x-strain on a number of different sized unirradiated graphite samples agreed with that measured by attached strain gauges. The discrepancy in the Young's modulus is most probably due to misalignment and friction associated with an unsuitable test fixture. An improved design, together with better modelling of the effects of the interaction between beam and test fixture is required.

The preliminary test carried out to measure the CTE has shown that ESPI has potential as a technique for irradiated graphite, although an improved heating arrangement to minimise temperature gradients is necessary.

Acknowledgements
The authors would like to thank British Energy Generation Ltd for permission to present this paper.

References
Kelly. B. T. (1981). Physics of Graphite, Applied Science Publishers, 122-124.
Timoshenko, S. P. and Goodier, J. N. (1970). Theory of Elasticity, McGraw-Hill, 113-122.

Pile Grade A Graphite – Poisson's Ratio

Mark Joyce[a] and Colin D. Elcoate[b]

[a]Frazer-Nash Consultancy Limited, 5th Floor, Malt Building, Wilderspool Business Park, Greenalls Avenue, Warrington, Cheshire, WA4 6HL.
[b]Frazer-Nash Consultancy Limited, 1 Trinity Street, College Green, Bristol, BS1 5TE, UK.
Email: m.joyce@fnc.co.uk

Abstract

Magnox graphite moderator bricks were extruded from Pile Grade A (PGA) graphite. The extrusion process tends to align the needle like crystallites with their basal planes parallel to the extrusion direction (along the bricks vertical axis). This alignment results in bricks that can be assumed to have orthotropic material properties, with an isotropic plane that is parallel to the extrusion direction. An understanding of internal stress and deformation of core components is essential in assessing component integrity, core geometry and consequently core safety. Poisson's ratio is a key input to the finite element analyses performed in support of reactor safety cases. However the evolution of this parameter with irradiation and/or oxidation is currently uncertain, leading to uncertainty in predictions that is typically dealt with by adopting a conservative approach. Typically Poisson's ratio is obtained from small sample tests either by monitoring the ratio of strains evolved during a loading cycle or via an ultrasonic technique. However both of these are inherently challenging in graphite due to the extremely small strains and inherently friable nature of the material. This paper describes progress from an on-going programme to derive Poisson's ratio via a novel non-contacting strain mapping approach. Indicative results, demonstrating the technique's applicability to both virgin nuclear graphite and a porous graphite material are presented. In addition, a review of existing data and numerical models, and their applicability to PGA graphite is presented.

Keywords
Poisson's ratio, Irradiation, Oxidation

INTRODUCTION

It is recognized that the available Poisson's ratio data for nuclear graphite is reasonably sparse; and particularly so with respect to its evolution with irradiation and / or oxidation. A good review of historical data is given by Preston (1998). However this data should be reassessed in conjunction with more recent data including that produced in the study by workers at the National Physical Laboratory (NPL), *i.e.* Lord *et al.* (2008). This overall review is required in order to support Magnox graphite core safety cases and is the subject of this paper.

MEASUREMENT

The quantity and consistency of data published on the Poisson's ratio of nuclear graphite is fairly limited. Elastic Poisson's ratio is inherently challenging to measure, being the ratio of two very small strains. Furthermore, historically, a wide range of graphite grades have been considered, which has lead to an inconsistent set of data. Generally two methods for measuring Poisson's ratio have been adopted; either the transverse and longitudinal strains are measured directly during a mechanical loading cycle, or the orthogonal stiffnesses are derived from the propagation of ultrasonic waves through the material. The latter method gives the dynamic Poisson's ratio, which may differ from the static value in the same way as Young's modulus, due to dependencies on the sample shape and ultrasonic wavelength, Maruyama (1982). Being directly related to the stiffness matrix, Poisson's ratio may be orientation dependent and will adopt six discrete values in a fully anisotropic material. Fortunately, some degree of isotropy may be reasonably assumed in nuclear graphites. Hence only a single value is typically quoted for nominally isotropic grades (*e.g.* Gilsocarbon), whilst three values are required to characterize transversely orthotropic grades (*e.g.* PGA).

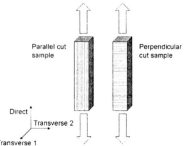

FIGURE 1: Sample orientations. Note for parallel cut samples, both transverse directions are perpendicular to extrusion. Whilst in perpendicular cut samples, one transverse direction is parallel and the other is perpendicular to extrusion.

To obtain the three discrete Poisson's ratio values required to characterize transversely orthotropic materials, samples are cut such that their direct loading axis is either parallel to or perpendicular to the extrusion direction, as shown in Figure 1; such that:

Parallel samples:
- Direct strain = Parallel to extrusion (Axial in brick sense);
- Transverse strain 1 = Perpendicular to extrusion (Hoop or Radial in brick sense);
- Transverse strain 2 = Perpendicular to extrusion (Radial or Hoop in brick sense).

Perpendicular samples:
- Direct strain = Perpendicular to extrusion (Hoop or Radial in brick sense);
- Transverse strain 1 = Perpendicular to extrusion (Radial or Hoop in brick sense);
- Transverse strain 2 = Parallel to extrusion (Axial in brick sense).

If the assumption of transverse orthotropy is correct, an identical Poisson's ratio should be measured in parallel cut samples irrespective of which transverse strain is measured. However two unique values should be observed in perpendicular samples, since the two transverse strains will be different.

VIRGIN VALUES

This section summarizes the available data on measured Poisson's ratios across a range of nuclear graphite grades in the un-irradiated condition. To allow comparison, the reported ranges for PGA are collated in Table 1.

Losty and Orchard (1961) conducted a series of tests on samples cut from a single PGA graphite brick. Samples were cut both parallel and perpendicular to the extrusion direction and tested in either compression or tension. Poisson's ratio values were obtained at a single stress level in tension and at four stress levels in compression by directly measuring the longitudinal and both transverse strains. The following range of results was obtained in compression:

- Parallel samples – transverse strain in hoop direction: 0.094 – 0.107.
- Parallel samples – transverse strain in radial direction: 0.070 – 0.079.
- Perpendicular samples – transverse strain in perpendicular direction: 0.124 - 0.146.
- Perpendicular samples – transverse strain in parallel direction: 0.054 – 0.057.

Whilst in tension:

- Parallel samples – transverse strain in hoop direction: 0.116.
- Parallel samples – transverse strain in radial direction: 0.112.
- Perpendicular samples – transverse strain in perpendicular direction: 0.156.
- Perpendicular samples – transverse strain in parallel direction: 0.07.

Brocklehurst and Lynam (1965) performed similar compression tests on both virgin PGA and on two prototypic Gilsocarbon grades termed P13 and P15. The following results were obtained for PGA samples:

- Parallel samples – transverse strain direction uncontrolled: 0.09 – 0.15.
- Perpendicular samples – transverse strain in perpendicular direction: 0.10.
- Perpendicular samples – transverse strain in parallel direction: 0.05.

Table 1: Collated reported Poisson's ratio values for un-irradiated PGA graphite.
* No specific reference given, but reported in Marsden and Wilkes (1998)

Sample Orientation	Losty & Orchard (1961)	Brocklehurst & Lynam (1965)	NPL (DIC) Lord *et al.* (2008)	Kelly*	Warner (1984)
Parallel	0.07 – 0.12	0.09 – 0.15	0.08 ± 0.02	0.05	0.06 – 0.11
Perpendicular, transverse perpendicular	0.12 – 0.16	0.10	0.12 ± 0.03	0.12	0.10 – 0.16
Perpendicular, transverse parallel	0.05 – 0.07	0.05	0.05 ± 0.01	0.12	0.10 – 0.16

In contrast, higher Poisson's ratios of 0.16 – 0.17 were obtained from the nominally isotropic Gilsocarbon samples. It should be noted that these samples were obtained from early production trials with bricks being produced by extrusion, Hanstock (1964), hence they may be more anisotropic and not directly comparable with the later moulded production materials.

In his review of Poisson's ratio, Preston (1988) reports a series of measurements on un-irradiated production Gilsocarbon moderator materials. The majority of measurements were obtained from samples of Heysham B / Torness production material (termed SI grade) and showed a degree of hysteresis; an average Poisson's ratio of 0.22 was measured during loading compared to 0.19 during unloading. Also reported are a series of measurements on other Gilsocarbon grades, these are referenced by Preston as un-published data by R.G.

Brown, these are of interest since many were also subject to irradiation and will be discussed in a subsequent section.

The following were reported:

- NA, a pre-production moulded trial material. Poisson's ratio: 0.23 - 0.25.
- NP, a fuel sleeve trial material. Poisson's ratio: 0.19 - 0.21.
- NY, Heysham A moderator production grade. Poisson's ratio: 0.19 - 0.20.
- NI, Dungeness B moderator production grade. Poisson's ratio: 0.16 - 0.21.

Recently NPL have completed an extensive study on the Poisson's ratio of both virgin PGA and Gilsocarbon. Over 200 compression tests were completed and Poisson's ratio obtained either using an ultrasonic technique (US) or from full field strain measurements made using Digital Image Correlation (DIC) during a mechanical loading cycle. The following results were obtained for PGA:

- Parallel samples where the transverse strain was measured in the hoop or radial direction. Average Poisson's ratios of 0.08 and 0.12 were measured using the DIC and US techniques respectively.
- Perpendicular samples where the transverse strain was measured in the perpendicular direction. Average Poisson's ratios of 0.12 and 0.21 were measured using the DIC and US techniques respectively.
- Perpendicular samples where the transverse strain was measured in the parallel direction. Average Poisson's ratios of 0.05 and 0.06 were measured using the DIC and US techniques respectively.

The Gilsocarbon samples were assumed to be isotropic and hence only a single value of Poisson's ratio was obtained from each technique; these were 0.19 and 0.23 using DIC and US, respectively.

EFFECT OF IRRADIATION

The effect of non-oxidising irradiation on the evolution of Poisson's ratio has been considered by several authors, their findings are summarized in this section. Brocklehurst and Lynam (1965) report the results of two sets of measurements on irradiated samples of PGA, which are shown in Figure 2. Unstressed samples in the parallel orientation were irradiated at 150°C, whilst samples cut in the perpendicular direction were irradiated at 200-250°C under an unspecified compressive stress. They found that Poisson's ratio appeared to increase with dose at both irradiation temperatures, with a far greater rise indicated at the lower temperature. The authors describe the sample irradiated at the lower temperatures as having become barrelled during irradiation. At this temperature, the dimensional change rate is likely to be relatively high, Kelly (1981), and hence any unintended flux or thermal gradient may result in sample distortion. Evidence for this behaviour is given by the apparent dependency of Poisson's ratio on the position at which transverse strain was monitored.

Taylor *et al.* (1967) report the results of a series of tests on nominally isotropic petroleum coke graphites, which are shown in Figure 3. As with the previously described study these were again irradiated at 150°C, but unlike the PGA sample considered by Brocklehurst and Lynam; after a modest rise at low dose, a decreasing trend was observed. Both the previous studies consider irradiation to only a low dose ($<15\times10^{20}$ n/cm^2). A small amount of data is available for Gilsocarbon samples irradiated to much higher dose ($>150\times10^{20}$ n/cm^2) and at a higher temperature of 450°C; this data is reported by Preston (1988) and described as unpublished data by R.G. Brown. The results, shown in Figure 4, appear quite scattered with the results denoted NA showing a strong rising trend, whilst those from NY and NX samples appear to decrease slightly.

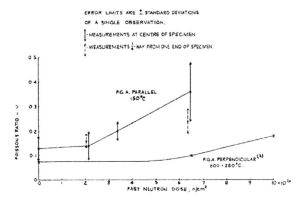

FIGURE 2: Effect of low temperature irradiation on the Poisson's ratio of PGA graphite, (Brocklehurst and Lynam, 1965).

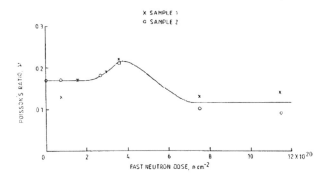

FIGURE 3: Effect of low temperature irradiation on the Poisson's ratio of isotropic graphite, replicated from Preston (1988), based on data from Taylor *et al* (1967).

As a final note, Shtrombakh *et al.* (2007) give values of Poisson's ratio for the GR-280 graphite grade employed in RBMK reactors. These are given as 0.16 in the virgin condition and 0.20 post-irradiation. Whilst several of these grades exhibit anisotropy in the virgin un-irradiated condition, unfortunately no data is available to characterize the evolution of multiple orientation dependent Poisson's ratios with dose.

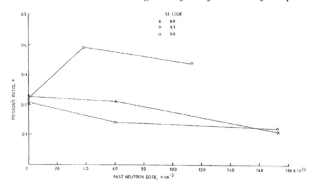

FIGURE 4: Effect of high dose irradiation at 450°C on the Poisson's ratios of selected
Gilsocarbon graphite grades, replicated from Preston (1988).

EFFECT OF OXIDATION

No Poisson's ratio measurements appear to be available from samples irradiated in an
oxidising environment. However some information may be inferred from the behaviour of
other materials since the role of porosity on Poisson's ratio has been well studied in ceramics
and a number of models considered. Figure 5 shows the results of an overlapping pore model
which suggests that as porosity increases Poisson's ratio tends to a single value independent
of the solid (zero porosity) value, Roberts and Garboczi (2000).

A further modelling study was conducted by Berre (2007) who produced finite element
models from X-ray tomographic reconstructions of thermally oxidized Gilsocarbon samples.
His results are shown in Figure 6 which shows that predicted Poisson's ratio remains
reasonably constant until a porosity fraction of approximately 0.3, before becoming more
scattered and generally decreasing. However the validity of these models at high porosity
values is questionable, since the pore size becomes increasingly comparable with the overall
model dimension.

NPL were able to make Poisson's ratio measurements on isotropic filter graphite (PG-25).
This has a low density of 1.08 g/cm^3 (equivalent to PGA at approximately 40% weight loss).
Poisson's ratio values of 0.17 and 0.16 were measured using the DIC and US techniques,
respectively.

The NX series of data points referred to previously, were obtained from a Gilsocarbon
reflector grade material. This is un-impregnated and has a significantly lower density than
the equivalent production moderator grades. Edwards (1973) (1.62 g/cm^3 c.f. 1.82 g/cm^3
typical of production Gilsocarbon). Its virgin Poisson's ratio is reported by Preston as 0.21 -
0.23, which is comparable with the upper range of the production moderator grades (NI, NY
and SI) 0.16 – 0.22. Furthermore, the irradiation responses of the NX and NY grades also
appear comparable despite their differing porosity fraction as shown in Figure 4.

Murayama *et al.* (1995) report the behaviour of a number of isotropic graphite grades after
irradiation to very high dose (up to 286×10^{20} n/cm^2) at 550-600°C. Poisson's ratio values
were obtained using an ultrasonic technique and are shown plotted against the inferred
porosity fraction for each sample in Figure 7. The irradiated results are higher and

significantly more consistent than the scattered virgin results. Interestingly, whilst the virgin values show little trend with increasing porosity, a slight decreasing trend is evident post-irradiation.

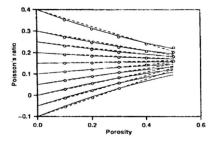

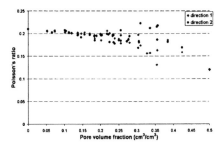

Figure 5: Modelled effect of porosity fraction on Poisson's ratio for a range of solid Poisson's ratios (zero porosity fraction), replicated from Roberts and Garboczi (2000).

Figure 6: Modelled Poisson's ratio in thermally oxidized Gilsocarbon as a function of porosity, replicated from Berre (2007).

DISCUSSION

The Poisson's ratio results obtained in early studies are similar to the values obtained from the more recent and extensive project by NPL; and are collated in Table 1. From these reports, there is compelling evidence for 3 unique values of Poisson's ratio in virgin PGA, hence it is intriguing why only two values are quoted by both Marsden and Wilkes (1998) and Warner (1984). However it is conservative in stress analyses to set both perpendicular Poisson's ratios to the higher measured value and furthermore Geering (2005) showed that, hoop stress and hence the potential for axial bore cracking is largely insensitive to the perpendicular Poisson's ratio value. Whilst Poisson's ratio measurements have been made on irradiated PGA samples, they appear inconsistent. The significant rise noted by Brocklehurst and Lynam (1965) after irradiation at 150°C was highly dependent on sample measurement method; hence it is suggested that this sample became distorted during irradiation. Furthermore it is not supported by the same author's results at 200°C which showed a far more modest rise. Whilst this latter sample was loaded during irradiation, there is now little evidence to suggest this should affect the elastic Poisson's ratio and hence the result appears valid.

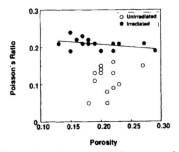

Figure 7: Poisson's ratio of a range of isotropic graphite prior to and post irradiation, replicated from Kennedy *et al.* (1977).

The majority of Poisson's ratio results in Gilsocarbon after irradiation at 450°C show a slight decreasing trend with dose, albeit from a higher virgin value than observed in PGA. The exceptions are two results obtained from samples of a pre-production Gilsocarbon. Previous work suggests that this grade is inherently more variable, particularly in its dimensional change response, and hence these results should be treated with caution, Joyce (2007). Interestingly there seems to be little difference between the irradiation response of the production moderator and reflector grades despite their significantly different porosity fractions. Whilst this cannot be used as absolute proof of no oxidation dependency; when taken together with the measurements of Maruyama *et al* and the model studies of Roberts and Garboczi (2000) and Berre (2007), it does suggest no first order correlation between porosity fraction and Poisson's ratio.

The results of Murayama *et al.* (1982) also suggest that irradiation tends to homogenise the Poisson's ratios of grades exhibiting low virgin values, leading to a more consistent value in the range 0.18 to 0.24 post irradiation. Unfortunately the doses considered by Murayama *et al.* (1982) are very high and hence no indication of the rate of saturation can be obtained. However, the magnitude of the increase is consistent with that observed at low dose by Brocklehurst and Lynam (1965) at 200°C, and that reported by Shtrombakh *et al.* (2007) for GR-280, used in RBMKs. The low temperature data appears to indicate that the process is relatively rapid. However, the mechanism is likely to be linked to the closure of accommodation porosity by crystalline shrinkage and hence is highly sensitive to temperature; the shrinkage rate being slower at higher temperatures.

It is interesting to note that in general, grades exhibiting high virgin Poisson's ratio values tend to decrease with irradiation, whilst grades with low virgin values appear to increase. It is therefore proposed that the studies which produced rising trends with dose are more likely to be representative of the behaviour of PGA. Finally it should be noted that changes in Poisson's ratio with dose will affect the measured dynamic Young's modulus values. Payne (1996) gives the correlation between ultrasonic wave velocity and Young's modulus as a function of orthotropic Poisson's ratios. Failure to correct for a rise in Poisson's ratio with dose will lead to a systematic over-prediction of measured dynamic modulus, the magnitude of which is a function of all three Poisson's ratio values.

CONCLUSIONS

The following conclusions were drawn from this study:

1. Overall, whilst there is good evidence for Poisson's ratio exhibiting dose sensitivity, it is suggested that in graphite exhibiting low virgin values this is a modest rise saturating at a value in the range 0.18 to 0.24. However, this assessment is largely made using data from isotropic graphite at very high irradiation dose.
2. Little evidence was found for any strong trend with weight loss, although what does exist suggests a weak decreasing trend.
3. It is important to consider the effect of Poisson's ratio on the value of dynamic Young's modulus when predicting stresses and strains in graphite components.

References

Berre, C. (2007). Microstructrual modelling of nuclear graphite using x-ray microtomography data. PhD thesis. University of Manchester.

Brocklehurst, J. E. and Lynam, J. T. (1965). Poisson's ratio of reactor graphite and the effect of irradiation. TRG report 901(C).

Brocklehurst, J. E. and Kelly, B. T. (1989). A review of irradiation induced creep in graphite under CAGR conditions. ND-R-1406 (S) June 1989.

Davies, M. A. and Bradford, M. R. (2007). The development of a revised model for irradiation induced creep of graphite: progress statement at September 2007. DAO/REP/JIEC/012/AGR/07. Presented to Graphite Core Committee. GCC/P(07) 017.

Edwards, J. A. (1973). Results from irradiation of graphite in DFR. Part 19 (199-29). UKAEA TRG memorandum 6362 (S).

Geering, P. J. (2005). Pile Grade A Graphite Poisson's Ratio Sensitivity Study. RS/E&TS/GEN/REP/0084/05.

Hanstock, R. F. (1964). The graphite physics programme. TRG memorandum 2266 (C).

Joyce, M. (2007). Forward predictions of graphite properties and weight loss. Tuning potential of revised graphite material model. FNC 33629-017/33338R, Issue 1 Oct 2007. GCC/P(07) 015.

Kelly, B. T. (1981). Physics of graphite. Applied science publishers, Barking, UK 1981.

Lord, J., Lodeiro, M., Klimaytys, G. and Jiang, J. (2008). Experimental study into the Poisson's ratio of graphite – Interim report. MEN/EWST/GEN/REP/0007/08, Issue 1. January 2008.

Lord, J., Lodeiro, M., Klimaytys, G., Morrell, R. and Jiang, J. (2008). Experimental study into the Poisson's ratio of graphite – final report. MEN/EWST/GEN/REP/0032/08, Issue 1, May 2008.

Losty, H. H. W. and Orchard, J. S. (1961). The strength of graphite. Proc. 5th Carbon Conf. 1, 519-32. Pergamon Press, London.

Marsden, B. J. and Wilkes, M. A. (1998). The effect of irradiation damage and radiolytic oxidation on PGA graphite. CSDMC/P133/A/1 Apr. 1998.

Maruyama, T., Katsumizu, K., Suzuki, H., Eto, M. and Oku. T. (1982). Measurement of elastic moduli of nuclear graphite. International symposium on carbon: New processing and new applications. Toyomashi Japan, 236.

Payne, J. F. B. (1996). On the relation between ultrasonic wave speed and elastic properties of an anisotropic solid. TE/GEN/REP/0101/96. August 1996.

Preston, S. D. (1988). Elastic Poisson's ratio of nuclear graphites – a review. ND-R-1490 (S) Oct 1988.

Roberts, A. P. and Garboczi, E. R. (2000). Elastic properties of model porous ceramics, *J. Amer. Ceram. Soc.*, 83, 3041-48.

Shtrombakh, Y., Platonov, P., Chugunov, O., Manevsky, V. and Alekseev, V. (2007). Report 2. Task A4 – Cross code comparison. Part 1, Moscow.

Taylor, R., Brown, R. G., Gilchrist, K., Hall, E., Hodds, A. T., Kelly, B. T. and Morris, F. (1967). The mechanical properties of reactor graphite, *Carbon*, 5, 519-531.

Warner, D. R. T. (1984). A review of PGA graphite physical property data. SWR/SSD/0347/R/84. May 1984.

Derivation of Dimensional Change Behaviour from Irradiated Brick Geometry

John F. B. Payne

National Nuclear Laboratory, Building 102(B), Stonehouse Park, Bristol Road, Stonehouse, Gloucestershire, UK, GL10 3UT
Email: john.b.payne@nnl.co.uk

Abstract

Dimensional change data are needed to predict the strain and stress in graphite reactor core components, and hence, to predict the probability of failure of such components. The stresses and strains in graphite bricks depend on gradients of dimensional change in the brick. This causes difficulty because MTR data is typically available only for dimensional change at particular values of fast neutron fluence and temperature. MTR data therefore have to be fitted to some plausible dependence on temperature and fast neutron fluence, which then has to be differentiated once or twice with respect to fluence and/or temperature to obtain the parameters governing brick stress and strain. These differentiations greatly magnify errors. This paper investigates how changes in the shape of a graphite reactor core brick in service may be used to deduce the gradients of perpendicular dimensional change in the brick, in addition to gradients of parallel dimensional change.

Keywords

Dimensional change, Reactor measurements, Core component stresses

INTRODUCTION

Fast neutron irradiation of graphite causes its dimensions to change. It is difficult to obtain accurate derivatives of graphite dimensional change with respect to temperature and fluence from Materials Test Reactor data. These derivatives are needed to predict the irradiation induced stresses in graphite components.

The PGA graphite used in Magnox reactors is anisotropic: the change in its dimensions parallel to the direction in which it was extruded during production ("parallel dimensional change") differs from the change in its dimensions perpendicular to the extrusion direction ("perpendicular dimensional change"). Perpendicular dimensional change is important to the demonstration of a low probability of axial brick cracking. This paper investigates how changes in the shape of a Magnox reactor graphite bricks in service may be used to deduce the gradients of perpendicular dimensional change in the brick, in addition to gradients of parallel dimensional change.

Changes in brick shape are commonly predicted from dimensional change by finite element analysis (FEA) as part of brick stress calculations. This work is concerned with the inverse problem of determining parallel and perpendicular dimensional change from brick shape. FEA is not very suitable for this inverse problem; analytical solutions that connect brick shape to dimensional change are required. The analytical solutions require some idealisations of the problem. The validity of derived dimensional change gradients can be checked by using FEA to calculate the brick shape from them.

THE BRICK MODEL & ITS GOVERNING EQUATION

The model is intended for Magnox reactor bricks, hence it is necessary to include graphite anisotropy. On the other hand, effects of high fast neutron fluence can be ignored. Changes to the model might be needed for reactors that are expected to reach high fluences. The model considers cylindrically symmetric changes in brick shape. Brick bowing is not cylindrically symmetric; it depends only on parallel dimensional change and is not considered in this paper, although it can be used to derive parallel dimensional change behaviour (Payne, 2005, unpublished).

The brick is idealised as a thin-walled hollow vertical cylinder and the forces acting on a vertical strip of the cylinder and their moments are considered. Figure 1 shows a vertical section of cylinder wall that is bent to a radius of curvature R by the moments of the forces acting on it. Let R_0 be the radius of curvature of the strip, after dimensional change has occurred, that makes the axial stress integrated through the wall thickness have zero moment; this implies that the average derivative of dimensional change parallel to z with respect to radial distance r is $1/R_0$.

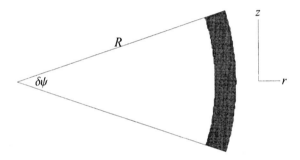

FIGURE 1: Vertical section of cylinder wall (z-axis vertical).

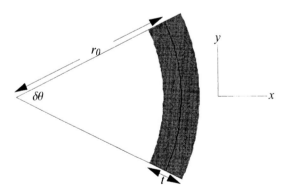

FIGURE 2: Horizontal section of sector of cylinder.

Now consider a horizontal section through part of the cylinder as shown in Figure 2. Let the cylinder mean radius (the radius to a point halfway through the cylinder wall) be r_m before dimensional change has occurred, and let r_0 be the mean radius of the cylinder after dimensional change has occurred, such that the elastic energy associated with hoop strain is a minimum. To a good approximation, the fractional perpendicular dimensional change is $\alpha_\perp = (r_0 - r_m)/r_m$. Let s be the cylinder radial deflection from the radius r_0. Then the elastic hoop strain in the cylinder is $s/r_0 \simeq s/r_m$ and is positive (tensile) if s is positive (that is to say cylinder radius has increased).

The details of the analysis are omitted, but, in essence, a balance of the moments of the axial stresses which change the curvature of a vertical strip of the cylinder from R_0 to R against the moments of the hoop stresses produced by the cylinder radial deflection, s, yields the equation

$$\frac{d^2}{dz^2}(s + r_0) + 1/R_0 = -\frac{12\,Y_\perp}{E_\parallel\,t^2\,r_0^2}\int_0^{z'} s(z)\,(z' - z)\,dz = -4\beta^4\int_0^{z'} s(z)\,(z' - z)\,dz \qquad [1]$$

where the left hand side is to be evaluated at $z = z'$, t is the cylinder wall thickness, $E_\parallel = Y_\parallel/(1 - v_{xz}\,v_{zx})$, $Y_\parallel$, $Y_\perp$ are Young's moduli in the vertical and hoop directions, respectively (parallel and perpendicular to the extrusion direction) and v_{xz}, v_{zx} are Poisson's ratios from the perpendicular to the parallel direction, and from the parallel direction to the perpendicular direction. The parameter β is defined by Equation [1[, and this equation can be converted to an ordinary differential equation by differentiating twice with respect to z', giving

$$\frac{d^4}{dz'^4}(s + r_0) + \frac{d^2}{dz'^2}\frac{1}{R_0} = -4\beta^4 s(z') \qquad [2]$$

Equation [2] is a necessary but not sufficient condition for $s(z)$ to satisfy Equation [1]: arbitrary constants in any solution to Equation [2] have to be chosen to satisfy Equation [1]. So far, the model has ignored graphite creep. For graphite bricks in Magnox reactors, creep strains are generally larger than elastic strains. Creep acts to reduce elastic strain at a rate proportional to the product of elastic strain and fast neutron flux. Thus, creep will reduce the elastic hoop and axial strains at each position in the brick by a nearly constant factor (since fast neutron fluence does not vary much with position). The stresses will therefore be reduced by a nearly constant factor and therefore will remain in equilibrium. The total strains will be virtually unchanged.

SOLUTIONS TO THE MODEL

Having derived the governing equation of the model (Equation [1]), it remains to find solutions to it. Mathematical details of the solution process are omitted. It is best to begin by considering the simplest case, which is the case of no axial variation of dimensional change.

Dimensional Change Independent of z — End Effects
If the cylinder is irradiated in a fast neutron flux that does not vary with z, then R_0, the mean radius of curvature of the strip at which strain is zero after dimensional change has occurred, and r_0, the mean radius of the cylinder such that the average hoop strain is zero after dimensional change has occurred, are both independent of z, so equation 2 simplifies to

$$\frac{d^4 s}{dz'^4} = -4\beta^4 s(z') \qquad [3]$$

Since fast neutron flux is independent of z the only radius variation occurs near the cylinder end, and it is simplest to consider a cylinder extending from $z = 0$ to $z = \infty$, for which

$$s(z) = \frac{e^{-\beta z}(\text{Sin}[\beta z] - \text{Cos}[\beta z])}{2R_0\beta^2} \qquad [4]$$

The radius variation near the cylinder end is shown in Figure 3 for dimensional change rate and other parameters typical of Magnox reactors. Note that the length scale, β^{-1}, is a small fraction of the brick length, which is ≈ 800 mm.

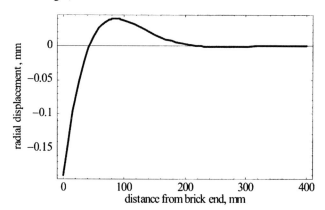

FIGURE 3: Radius Variation near Brick End.

The solution given by Equation [4] is essentially a generalization to an anisotropic material of the result given by Timoshenko & Goodier (1970). For the isotropic case, the solutions agree, once a typographical error in Timoshenko & Goodier is corrected.

Parallel Dimensional Change Varying with z

The axial flux shape in the reactor core often results in an approximately linear variation of parallel dimensional change rate with height z that produces a variation of

$$1/R_0(z) = 1/R_0(0) + bz$$

where b is a constant. The solution for this case is

$$s(z) = c_1 \, \text{Cosh}[\beta z]\,\text{Cos}[\beta z] + c_2 \, \text{Sinh}[\beta z]\,\text{Sin}[\beta z] + \\ c_3 \, \text{Cosh}[\beta z]\,\text{Sin}[\beta z] + c_4 \, \text{Sinh}[\beta z]\,\text{Cos}[\beta z] \qquad [5]$$

where z is vertical distance from the brick centreline.

The constants c_1, c_2, c_3, c_4 are determined by the brick height and the variation of parallel dimensional change with z, as follows:

$$c_1 = \frac{\text{Cosh}[\beta h]\,\text{Sin}[\beta h] - \text{Cos}[\beta h]\,\text{Sinh}[\beta h]}{R_0(0)\,\beta^2\,(\text{Sin}[2\beta h] + \text{Sinh}[2\beta h])}$$

$$c_2 = \frac{-\text{Cosh}[\beta h]\,\text{Sin}[\beta h] - \text{Cos}[\beta h]\,\text{Sinh}[\beta h]}{R_0(0)\,\beta^2\,(\text{Sin}[2\beta h] + \text{Sinh}[2\beta h])}$$

$$c_3 = \frac{b\,\text{Cosh}[\beta h]\,\text{Sin}[\beta h]\,(\beta h\,(\text{Cot}[\beta h] + \text{Tanh}[\beta h]) - 1)}{\beta^3\,(\text{Sin}[2\beta h] - \text{Sinh}[2\beta h])}$$

$$c_4 = \frac{b\,\text{Cosh}[\beta h]\,\text{Sin}[\beta h]\,(\beta h\,\text{Cot}[\beta h] - (\text{Cot}[\beta h] + \beta h)\,\text{Tanh}[\beta h])}{\beta^3\,(\text{Sin}[2\beta h] - \text{Sinh}[2\beta h])}$$

[6]

where h is the half brick height. The $s(z)$ variation with z calculated from Equations [5] and [6] is shown in Figure 4 based upon brick dimensions, fast neutron fluences and dimensional change rates that occur in Magnox reactors. The effect of the axially varying parallel dimensional change rate is to make the end effects asymmetric.

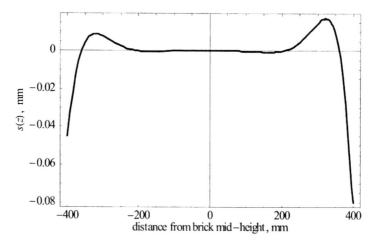

FIGURE 4: Deflection from zero hoop strain radius, due to linearly varying $1/R_0(z)$.

Perpendicular Dimensional Change varying with z

The solution given by Equations [5] and [6], which allows for approximately linear axial variation of fast neutron flux, takes no account of the effect of axial variation of fast neutron flux on perpendicular dimensional change. Approximately linear axial variation of fast neutron flux will result in approximately linear variation of r_0 with z. A linear variation of r_0 with z makes no contribution to $s(z)$ and the non linear contribution of r_0 with z in Magnox reactors can be shown to make a negligible contribution to $s(z)$. Consequently the solution for $s(z)$ reactors given by Equations [5] and [6] holds in this case. The brick end effects depend on parallel dimensional change, but away from the brick ends the brick radius is determined by perpendicular dimensional change, to a good approximation.

Figure 5 shows the predicted radius variation (solid line) and the measured brick internal radii for an Oldbury reactor brick (the dots). The raw measured brick internal radii data vary cyclically with height, with a wavelength of ~50 mm; this is thought to be either a result of the bore machining or a systematic measurement error.

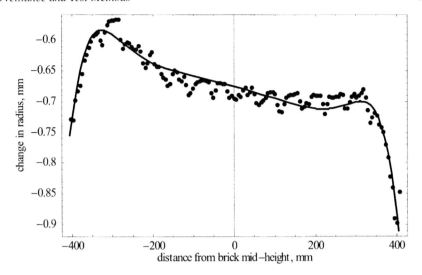

FIGURE 5: Change in brick radius with linearly varying $1/R_0(z)$ and $r_0(z)$.

The prediction and data agree reasonably well, bearing in mind that details of geometry (spigotting at brick ends) and details of flux variation are both ignored. However, at Oldbury, although one bore diameter decreases at the brick end as shown in Figure 5, the perpendicular bore diameter often increases at the brick end. This anomaly does not occur at Wylfa; at Wylfa, the end effects on perpendicular diameters are the same. The anomaly may arise because the weight of the bricks makes an additional contribution to brick distortion (Oldbury bricks rest on their end faces, whereas Wylfa bricks rest on the spigot). The anomaly is not considered further because the model only considers cylindrically symmetric brick distortions.

EFFECT OF LOCALISED FAST NEUTRON FLUX REDUCTION

The uranium (which is the fast neutron source) is not continuous up a Magnox reactor channel; it is present in separate fuel elements. The fast neutron flux reduction due to the gap between uranium bars will produce a local reduction in dimensional change, and corresponding local increases in r_0 and R_0. To estimate the effect of the gaps in the uranium, the increases in r_0 and R_0 are assumed to be given by

$$\frac{1}{R_0(z)} = \frac{1}{R_0(0)} - a_1 e^{-(k(z-z_0))^2} \qquad [7]$$

$$r_0(z) = r_0(0) + a_2 e^{-(k(z-z_0))^2} \qquad [8]$$

where k is a constant that depends on the size of the gap, z_0 is the position of the centre of the gap, and a_1, a_2 are constants which are usually positive. A decrease in fast neutron fluence produces a decrease in parallel dimensional change, and thus an increase in R_0. It also produces an increase in r_0 if the perpendicular dimensional change is a shrinkage. The choice of a Gaussian curve is a compromise between simplicity and a physically reasonable variation (for example, it would be simpler to assume a quadratic variation over a limited range, but this would introduce physically implausible slope discontinuities in r_0 and R_0). The Magnox fuel elements are longer than the bricks, so each brick has at most one uranium gap within it.

The details of the solution of Equation [1] with $R_0(z)$ and $r_0(z)$ by Equations [7] and [8] are omitted. The important result is that, to a good approximation, the increase in brick radius caused by the fast neutron flux dip due to the uranium gap is a direct reflection of the unconstrained change in radius associated with perpendicular dimensional change. The assumed form of the axial variation of $r_0(z)$ matches the observed variation well, as Figure 6 shows. The solid line is the sum of a linear variation of radius with height and the assumed variation due to the uranium gap of Equation [8] (the end effects given by Equations [5] and [6] are not included). As in the Oldbury data, there is cyclical variation of measured bore diameter with height, with a wavelength of ~50 mm.

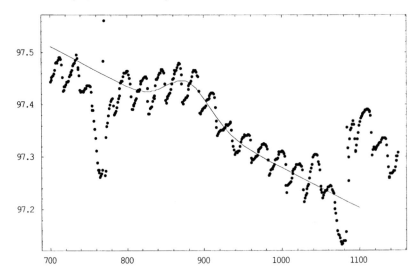

FIGURE 6: Data for Wylfa R2 example brick.

MEASURED BRICK GEOMETRY DATA REDUCTION

The solution for linearly varying flux (Equation [5]) can be combined with Equation [8] (and a term giving the linear variation of r_0 with z) to give a form for the axial variation of brick internal radius. The parameters in the solution can then be found by maximum-likelihood fitting to the brick internal radii measured in-reactor for each brick. The through wall gradient of parallel dimensional change and the axial variation of perpendicular dimensional change for the in-service condition of the brick can then be deduced from the fitted parameters.

CONCLUSIONS

The analysis given here shows how the gradients of parallel and perpendicular dimensional change in a brick can be deduced from the axial variation of the brick internal radius. Solutions have been obtained for various assumed axial variations of perpendicular dimensional change and through wall parallel dimensional change gradient. The assumed axial variations cover the cases that typically occur in practice. Using these solutions, the through wall gradient of parallel dimensional change in a brick can be deduced from the radius variation near the ends of an irradiated brick, and the variation of perpendicular dimensional change can be deduced from the radius variation elsewhere in the brick.

Acknowledgements
This paper is based on work done for Magnox North. Permission to publish this paper is gratefully acknowledged.

References
Timoshenko, S. P. and Goodier, J. N. (1970). Theory of Elasticity, 3rd Edition, McGraw-Hill.

Biaxial Testing: Appropriate for Mechanical Characterisation?

Gary D. Kipling, Andrew Easton and Gareth B. Neighbour

Dept. of Engineering, University of Hull, Cottingham Road, Hull, UK. HU6 7RX

Email: g.d.kipling@2005.hull.ac.uk

Abstract

Material properties are often described as being characteristic and are quoted as such irrespective of scale. However, although the testing of materials is sometimes assumed to be a mature field, there are issues related to the appropriateness of the test, the inherent suitability of the assumptions in determining the material property value, and indeed inherent microstructure of the material concerned. Quasi-brittle materials by their very character show deviation from a truly elastic material and so challenge some of the assumptions being made; and this is true for nuclear graphites. Polygranular graphite is used in Advanced Gas-cooled Reactors (AGRs) primarily as a means of providing moderation for the nuclear reaction, but also as a major structural component in the form of the core bricks. At present, in the nuclear industry, the prediction of when cracked graphite bricks will occur in a nuclear core is largely based on the measurement of mechanical properties from small samples, even though the volume of a typical brick is a factor of 10^4 greater than that of a typical flexural test sample. For polygranular graphites, many models to predict the probability of failure have been generated and these are usually related to a uniaxial value and indeed most tests conducted determine uniaxial values. If sample size restrictions apply, particularly for engineering ceramics, a biaxial stress geometry is sometimes used. This is especially true if the material is used in applications that impose multi-axial stress fields and so to some extent better resemble the engineering duty and reflect the real performance criterion. As an illustration, this paper will also discuss the evaluation and choice of different designs of biaxial test apparatus. Further, for the preferred biaxial testing system, results will be presented that demonstrates the issues discussed above and shows the complexities involved in the small-scale dependence of geometry upon strength.

Keywords

Biaxial stress, Strength, Mechanical characterisation

INTRODUCTION

The testing of materials is sometimes assumed to be a mature field. There are, however, issues relating to the appropriateness of the test, the inherent suitability of the assumptions in determining the material property value, and indeed inherent microstructure of the material concerned. Quasi-brittle materials by their very character show deviation from a truly elastic material and so challenge some of the assumptions being made. At present, in the nuclear industry, the prediction of when graphite bricks will crack in a nuclear core is largely based on the measurement of mechanical properties from small samples, even though the volume of a typical brick is a factor of 10^4 greater than that of a typical flexural test sample.

For polygranular graphites, many models to predict the probability of failure have been generated and these are usually related to a uniaxial value and indeed most tests conducted determine uniaxial values. If sample size restrictions apply, particularly for engineering ceramics, a biaxial stress geometry is sometimes used. This is especially true if the material is used in applications that impose multi-axial stress fields and so to some extent better resemble the engineering duty and reflect the real performance criterion. Thus, using a biaxial stress geometry is appealing in many respects, as it represents a more severe stress state than uniaxial stress and is accordingly better suited to conservative design basis. However, whilst biaxial testing has attractions there are additional complications which are the subject of this paper.

BIAXIAL TEST METHODS

There are many test configurations available to determine the biaxial strength of a material. In this work, an extensive literature review was performed searching for all types of biaxial test. After a brief evaluation, 11 candidate methods were identified:

- Diametral Compression Test (*e.g.* see Awaji *et al.* 1987)
- Ball on Three Ball (*e.g.* see Godfrey, 1985)
- Ring on Ring (*e.g.* see Fett *et al.* 2006)
- Ball on Ring (*e.g.* see Isgrò *et al.* 2003)
- Internal Pressure Tube / Ring Test (*e.g.* see Perreux and Suri, 1997)
- Bulge Test (*e.g.* see Imaninejad *et al.* 2004)
- Cruciform Test (*e.g.* see Welsh and Adams, 2002)
- Arcan Test (*e.g.* see Doyoyo and Mohr, 2003)
- Iosipescu Test (*e.g.* see Kumosa and Han, 1999)
- Cold-Spin Test (*e.g.* see Brückner-Foit *et al.* 1993)
- Scissor Arms Test (*e.g.* see Kumosa and Han, 1999)

A selection process was undertaken to determine the most appropriate test from the 11 candidate tests using a binary dominance method to weight each of the 17 identified selection criteria, including, for example, adaptability to different specimen geometries, potential for edge effects, suitability for brittle materials, *etc.* The most suitable test for the given attributes was found to be the ball-on-three-ball test. After selection, an experimental test programme was defined and the test apparatus was designed using SolidWorks and manufactured to fit existing high specification universal test machines (Lloyds EZ50).

Ball-on-Three-Ball Test Method
The ball-on-three-ball test apparatus comprises of a thin disc sample supported on three equally spaced ball bearings and held in position using alignment pins. The sample is loaded in the centre of the disc using another ball bearing, as illustrated in Figure 1. During testing, the bottom surface of the sample is subject to a biaxial tensile stress. Cheng *et al.* (2003) states that "*the crack extension initiates exclusively on the tensile free surface*" and also that "*compressive stresses normal to a crack will not cause fracture in brittle materials*". Fracture of biaxial samples tested using this method is therefore likely to be initiated at a crack or flaw on or near the surface of the material under tensile-tensile loading.

Analytical Solutions for 'Ball on Three Ball' Test
There are numerous analytical solutions for the ball-on-three-ball test method, each of which appears to be very different in nature. The most commonly used analytical solutions can be found in: Godfrey (1985); F394-78 Standard (1996); Ovri (2000); Higgs *et al.* (2000); Danzer *et al.* (2001); and Pagniano *et al.* (2005). An evaluation of these solutions was undertaken and is briefly described below with Table 1 defining the variables used in the analytical solutions presented herein.

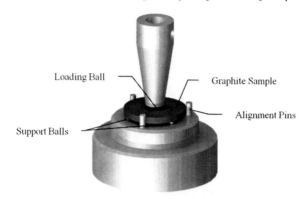

FIGURE 1: Rendered SolidWorks drawing of the Ball-on-Three-Ball test apparatus.

TABLE 1: Variables used in analytical solutions.

Variable type	Variable	Symbol	Variable type	Variable	Symbol
Applied Condition	Load	L		Diameter of ball	D_b
Material Geometry	Thickness	T	Apparatus Conditions	Poisson's Ratio (Ball)	v_b
	Radius	R_d		Young's Modulus (Ball)	E_b
Material Properties	Young's Modulus (Disc)	E_d		Support Radius	A
				Radius of ball	R_b
	Poisson's Ratio (Disc)	v_d	Contact Radius Approximations	Godfrey	R
				Westergaard	b

Investigation into the derivation of all these solutions revealed that they are similar in their manipulation of both material properties and experimental values. They are in essence approximated based upon 'cylindrical symmetrical thin-plate theory' for truly elastic materials (Kirstein and Woolley, 1966) which predicts an infinite stress amplitude opposite to the load transfer point; and they can all be related to the master equation (Marshall, 1980) for the biaxial stress, σ, such that:

$$\sigma = YL/T^2 \tag{1}$$

Our evaluation, however, has found that there are key differences regarding contact mechanics between the disc and the loading ball, and also the geometric factors used, Y. For example, Pagniano (2005), Equation [2], suggests that the radius of uniform loading at centre is equivalent to the radius of the loading ball.

$$\sigma = 3L \frac{(1+v_d)}{4\pi T^2}\left[1 + 2\ln\left(\frac{A}{R_b}\right) + \left(\frac{1-v_d}{1+v_d}\right)\left(1 - \frac{R_b^2}{2A^2}\right)\left(\frac{A^2}{R_d^2}\right)\right] \tag{2}$$

However, another solution, proposed by Godfrey (1985), Equation [3], includes an approximation for the contact radius between the loading ball and the disc described by Equation [4] using Hertzian theory (elastic body interaction) and so takes into account the material properties of the indenter and the disc.

$$\sigma = 0.4775\left(\frac{L}{T^2}\right)(1+v_d)\ln\left(\frac{A}{R}\right) + \frac{1}{2}(1+v_d) + \frac{0.25(1-v_d)(2A^2 - R^2)}{R_d^2} \qquad [3]$$

$$R = 0.721\left[\frac{LD_b(1-v_b^2)}{E_b} + \frac{(1-v_d^2)}{E_d}\right]^{\frac{1}{2}} \qquad [4]$$

Interestingly, solutions suggested by Higgs *et al.* (2000), Equation [5], and Danzer *et al.* (2001) also require an approximation for the contact radius. Both papers explain that Equation [4] is only valid for values of R which are larger than $1.7T$. Values of R smaller than $1.7T$ can be found by replacing the actual radius by an 'equivalent radius', b. An approximation for this 'equivalent radius' is given by Westergaard (1926), Equation [6].

$$\sigma = 3L\frac{(1+v_d)}{4\pi T^2}\left[1 + 2\ln\left(\frac{A}{R}\right) + \frac{(1-v_d)}{(1+v_d)}\left(1 - \frac{R^2}{2a^2}\right)\frac{A^2}{R_d}\right] \qquad [5]$$

$$b = \sqrt{1.6R^2 + T^2} - 0.675T \qquad [6]$$

Further, ASTM F394-78 Standard (1996) and Ovri (2000) do not use any contact radius approximations. Rather, these papers suggest the use of a hardened dowel to apply the load to the sample. The area of applied load is therefore stated as the radius of the dowel. In order to apply these solutions to the ball-on-three-ball method, it may be necessary to substitute an approximation for the contact radius. However, it should be noted that ASTM have now withdrawn this standard.

EVALUATION OF ANALYTICAL SOLUTIONS

For each of the most commonly used analytical solutions, an evaluation was undertaken using MathCAD for the test apparatus illustrated in Figure 1. For notionally the same applied load the solutions were shown to give rise to large differences in predicted biaxial strength as shown in Figure 2. This graph demonstrates the difference in calculated values for biaxial strength by imputing the same variables into the six solutions. Bizarrely, three analytical solutions yield a negative biaxial strength. Evidently these solutions cannot be valid for the conditions of the experiment. It is a likely that these solutions are applied to experiments using different materials or indeed samples of a very different geometry. Further modelling of the solutions using surface plots, illustrates the issue of scale when applying these solutions. Figure 3 shows that as the thickness of the sample increases, the sensitively to load decreases.

Figures 4a and 4b illustrate the issue regarding the contact approximation for the solutions. Figure 4a (solution from Godfrey (1985)) shows that this solution would predict a decrease is sensitivity as the thickness increases. The solution from Pagniano (2005) (Figure 4b) again predicts a decrease is sensitivity as the thickness increases. Additionally, however, when large balls are use to support the disc, a negative biaxial strength is calculated. Evaluation of the suitability of the solutions reveals that, for this investigation, only Godfrey (1985), Pagniano *et al.* (2005) and Higgs *et al.* (2000) could be applied to this test arrangement. Preliminary calculations using the solutions suggested by these three papers do yield encouraging results, albeit with concerns regarding the scale of the samples. The remaining three solutions are evidently not suitable for the testing of graphite using the suggested sample geometry.

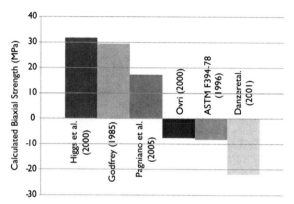

FIGURE 2: Biaxial strength of possible biaxial strength solutions.

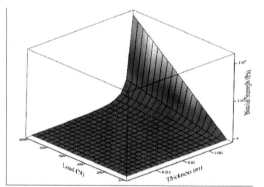

FIGURE 3: Relationship between load, sample thickness and biaxial strength (using Godfrey, 1985).

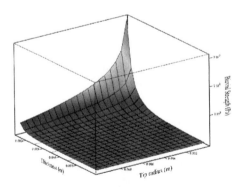

FIGURE 4a: Relationship between thickness, tip radius and biaxial strength (using Godfrey, 1985).

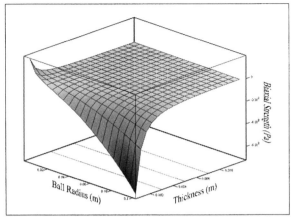

FIGURE 4b: Relationship between thickness, ball radius and biaxial strength (using Pagniano, 2005).

EXPERIMENTAL DETAILS

The material investigated is EY9 grade graphite. The material is supplied in bars of 25.4mm diameter to allow easy machining into suitably sized specimens. Typical properties of the graphite are summarised in Table 2.

TABLE 2: Typical properties of investigated graphite.

Property	EY9 (Williams *et al.* 1993)
Porosity (%)	17
Elastic Modulus (GPa)	13.1
Density (kg/m^3)	1677.0 (measured)
Tensile Strength (MPa)	14.11 (measured)
Compressive Strength (MPa)	51.0

Samples of EY9 grade graphite were prepared in thicknesses of 1, 2, 3, 4, 5 and 8 mm with a common diameter of 25mm. These samples were machined using a lathe. The material was first cut to the correct diameter. The individual specimens were then cut at the correct thickness using a 'parting off tool'. This process ensures that the faces of the sample parallel and the thickness is as accurate as possible. The specimens are then smoothed using fine silicon carbide paper.

The three ball bearing supports and loading ball were held in the correct position using grease. The sample was positioned using the pins on the base. The loading ball was lowered into a starting position just above the surface of the sample. A compression test was started, using a speed of 0.5mm/minute. The test continued until the specimen fractured, the load at this point (maximum load) is the value required to calculate the biaxial strength of the sample. NEXYGEN MT materials testing software was used to display and record properties of each test.

Determination of the materials uniaxial strength was achieved using the brittle ring test. This test comprises a ring sample which is compressed between two loading plates until fracture. Kennedy (1993) suggests that the brittle ring test results yield "*realistic estimates of strength for actual components*". The solution suggested by Kennedy to calculate the uniaxial strength of a sample is;

$$\sigma_{nom} = \frac{3P(a+b)}{\pi h(b-a)^2} \qquad\qquad \sigma_{max} = \sigma_{nom}K_t \qquad\qquad [6a, 6b]$$

Where, P is the applied load, a is the inner radius of the sample, b is the outer radius of the sample, h is the thickness of the sample and K_t is a value read from a plot in Petterson (1953). Brittle ring samples were placed in the testing machine between two loading plates before a compression test was conducted at a speed of 0.5 mm/min. The test continued until fracture of the specimen occurred; the maximum load at this point was recorded.

RESULTS AND DISCUSSION

The biaxial strength of EY9 graphite was calculated using three equations, all of which yielded different relationships between biaxial strength and sample thickness (shown in Figure 5). Higgs *et al.*, Godfrey and Pagniano *et al.* showed an increase in biaxial strength as the thickness of the specimen increased, however, the Pagniano equation did result in lower strength values than Godfrey. Godfrey (WG) (using Westergaard's approximation for contact radius) showed a roughly consistent value for biaxial strength as the thickness increased as did Higgs *et al.* (WG). Generally, the mean biaxial strength of a material tends to be lower than the uniaxial strength. Brocklehurst (1977), states that the biaxial strength of polygranular graphite is approximately 80 – 85% of the uniaxial strength. Brocklehurst goes on to explain that "*flaw mechanism of failure is that under a under a biaxial stress there is a greater chance of the larger flaws being critically oriented to a critical stress*".

Test data using the ball on three ball method suggest that the biaxial strength is generally higher than the uniaxial strength. This relationship contradicts the general relationship stated by Brocklehurst, that the biaxial strength is lower than the uniaxial strength.

The relationship shown by the five equations does not adhere to the theory of brittle fracture, whereby an increase in volume will result in a decrease in strength. This effect is attributed to the probability that a larger sample will be likely to contain critical flaws. A possible explanation for this effect is that as the thickness of the specimen increases, the sample experiences more compressive force from loading. Whilst the volume of the specimen is increased, the tensile area remains relatively constant. If this theory is correct, it could be concluded that the ball on three ball test method is not valid for samples which are subject to large amounts of compressive force. There was evidence of localised plastic deformation, caused by large compressive stresses at the point of loaded. This was particularly evident in the 8 mm thick samples which shows Hertzian cone cracking, resulting in a 'cup and cone' feature at the point of contact between ball and disc.

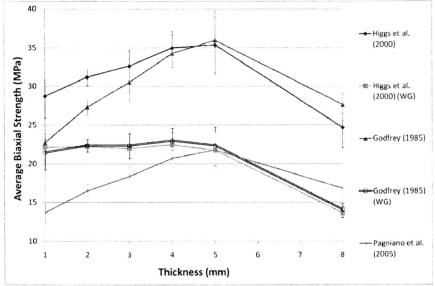

FIGURE 5: Calculated biaxial strength of EY9 grade graphite at varying thicknesses.

A total of 10 brittle ring tests were undertaken for EY9 grade graphite. The average uniaxial strength was calculated as 14.44 with a standard deviation of 1.10. Brocklehurst states that the compressive strength of graphite is typically 3 to 4 times higher than the tensile strength. EY9 Grade graphite has a compressive strength of 51.02 MPa (Williams *et al.* 1993), while the average tensile strength was calculated at 14.44 MPa. Using the aforementioned relationship yields an acceptable value of 3.53. The approach undertaken in this paper has also be applied to other materials such as POCO graphite (Easton, 2007) and Duratec 750 (Kipling, 2008). The results of these tests support the findings of this paper.

CONCLUSIONS

The biaxial strength values for EY9 grade graphite were higher than would have been predicted based on the materials uniaxial strength. The relationship between biaxial strength and sample thickness was also contrary to the expected trend. These results can be attributed to the stress distribution of the material. Finite element analysis on the 'ball on three ball' showed that the region of maximum tensile stress is very small, almost pin like Higgs *et al.* (2000). This effect also increases the effect of surface defects, since a flaw around the area of highest stress would have a large effect on the calculated strength of the material.

Disregarding the 8 mm samples, which appears are subject to a more complex stress field during loading, the most appropriate results are generated using the solutions suggested by Godfrey and Higgs and the contact approximation from Westergaard. The values for biaxial strength are higher than would be expected taking into consideration the relationship stated in Brocklehurst. These solutions do however yield the lowest calculated values for biaxial strength and are closest to the expected relationship between uniaxial and biaxial strength.

Analytical assumptions are likely to contribute to the higher calculated stress. The solutions are primarily used to calculate the biaxial strength of ceramics such as alumina, silicon nitride and glass. In each case the material being tested is brittle. Whereas graphite is commonly regarded as quasi brittle and it is unlikely that the analytical assumptions will take account of this factor.

References
ASTM F394-78 (Reapproved 1996). Standard test method for biaxial flexure strength of ceramic substrates.
Adams, D. F. and Welsh, J. S. (2002). An experimental investigation of the biaxial strength of IM6/3501-6 carbon/epoxy cross ply laminates using cruciform specimens. *Composites: Part A, 33*, 829-839.
Awaji, H., Sato, S., Kawamata, H., Kurumada, A., and Oku, T. (1987). Fracture criteria of reactor graphite under multiaxial stresses. *Nuclear Engineering and Design, 103*, 291-300.
Brückner-Foit, A., Fett, T., Munz, D., and Schirmer, K. (1997). Discrimination of multiaxiality criteria with the Brazilian disc test. *Journal of the European Ceramic Society, 17*, [5], 689-696.
Cheng, M., Chen, W., and Sridhar, K. R. (2003). Biaxial flexural strength distribution of thin ceramic substrates with surface defects. *Journal of Solids and Structures, 40*, 2249–2266.
Brocklehurst, J. E. (1977). Fracture in Polycrystalline Graphite. *Chemistry and Physics of Carbon*, [13], 145-279.
Danzer, R., Supancic, P., and Börger, A. (2001). The ball on three balls test for strength testing of brittle discs: stress distribution in the disc. *Journal of the European Ceramics Society, 22*, 1425-1436.
Doyoyo, M., and Mohr, D. (2003). Microstructural response of aluminium honeycomb to combined out-of-plane loading. *Mechanics of Materials, 35*, 865-876.
Easton, A. (2007). Determination of the effect of sample geometry upon the biaxial strength of engineering materials. BEng Thesis, University of Hull.
Fett, T., Rizzi, G., Ernst, E., Müller, R., and Oberacker, R. (2006). A 3-balls-on-3-balls strength test for ceramic disks. *Journal of the European Ceramic Society, 27*, [1], 1-12.
Feilzer, A. J., Isgrò, G., Pallav, P., and van der Zel, J. M. (2003). The influence of the veneering porcelain and different surface treatments on the biaxial flexural strength of a heat-pressed ceramic. *The Journal of Prosthetic Dentistry, 90*, [5], 465-473.
Godfrey, D. J. (1985). Fabrication, formulation, mechanical properties, and oxidation of sintered Si_3N_4 ceramics using disk specimens. *Materials Science and Technology, 1*, 510-515.
Han, Y. and Kumosa, M. (1999). Non-linear finite-element analysis of Iosipescu specimens. *Composites Science and Technologies, 59*, 561-573.
Higgs, R. J. E. D., Lucksanasombool, P., Higgs, W. A. J., and Swain M. V. (2000). Evaluating acrylic and glass-ionomer cement strength using the biaxial flexure test. *Biomaterials, 22*, 1583-1590.
Imaninejad, M., Subhash, G., Loukus, A. (2004). Experimental and numerical investigation of free-bulge formation during hydroforming of aluminum extrusions. *Journal of Materials Processing Technology, 147*, [2], 247-254.
Kirstein, A. F. and Woolley, R. M., (1967). Symmetrical bending of thin circular elastic plates on equally spaced point supports. *Journal of Research of the National Bureau of Standards, 71C*, 1–10.
Kipling, G. D. (2008). Determination of the effect of sample geometry upon the biaxial strength of engineering materials, BEng Thesis, University of Hull.
Kennedy, C. R. (1993). The brittle-ring test for graphite. *Carbon, 31*, [3], 519-528.
Ovri, J. E. O. (2000). A parametric study of the biaxial strength test for brittle materials. *Materials Chemistry and Physics, 66*, 1–5.
Pagniano, R. P., Seghi, R. R., Rosential, S. F., Wang, R. and Katsube, N. (2005). The effect of a layer of resin luting agent on the biaxial flexure strength of two all-ceramic systems. *J. of Prosthetic Dentistry*, 459-466.
Perreux, D. and Suri, C. (1997). A study of the coupling between the phenomena of water absorption and damage in glass / epoxy composite pipes. *Composites Science and Technology, 57*, 1403-1413.
Petterson, R. E. (1953). Stress concentration design factors. Wiley, New York, 135.

Part D

Prediction of Component Performance

The Development of a Microstructural Model to Evaluate the Irradiation-Induced Property Changes in IG-110 Graphite using X-ray Tomography

Junya Sumita[a], Taiju Shibata[a], Eiji Kunimoto[a], Graham Hall[b], Barry J. Marsden[b] and Kazuhiro Sawa[a]

[a]Japan Atomic Energy Agency, 4002, Narita-cho, Oarai-machi, Higashiibaraki-gun, Ibaraki-ken, 311-1394, Japan.
[b]The University of Manchester, P.O. Box 88, Manchester M60 1QD, United Kingdom.
Email: sumita.junya@jaea.go.jp

Abstract

IG-110 fine grain graphite is one of the candidate materials for core components in the Generation IV VHTR. In order to develop a microstructural / property change model for IG-110 graphite, the relationship between changes in the bulk properties and the microstructure of IG-110 graphite have been investigated. In this study, the microstructure of IG-110 graphite was modified by thermal oxidation and the resulting change in electrical resistivity was measured. Improvements in the resolution of X-ray tomography now allow the fine grain structure of IG-110 graphite to be successfully imaged. The following results were obtained. (1) IG-110 graphite samples were oxidized uniformly in air at 500°C and the relationship between electrical conductivity and burn-off (porosity) of IG-110 graphite obtained. (2) Three-dimensional images based on high resolution X-ray tomography were used to describe the microstructure change in IG-110 graphite. This study has shown that it should be possible to use the methodology presented in this paper to predict changes in the bulk properties of IG-110 graphite due to fast neutron irradiation.

Keywords

IG-110 graphite, Electrical conductivity, X-ray tomography

INTRODUCTION

The High Temperature Gas-Cooled Reactor (HTGR) has received much attention because the coolant helium gas can reach high temperatures at the reactor outlet and it has an inherent safety feature due to the large heat capacity of graphite components in the core. The High Temperature Engineering Test reactor (HTTR), constructed by Japan Atomic Energy Agency (JAEA) is the first HTGR in Japan. The Very High Temperature Reactor (VHTR) is one of the most promising candidates for the Generation IV Nuclear Energy System. The VHTR is capable of providing high temperature helium-gas at approximately 1000°C for electricity supply and process heat, e.g. for hydrogen production, as proposed in the Generation IV International Forum (US DOE, 2002).

IG-110 fine grain, isostatically pressed isotropic graphite (TOYO TANSO Co., Ltd.), is used for the core components in the HTTR and is also used in its unpurified form, IG-11 graphite, as the reflector in the 10MW High Temperature Gas-cooled Reactor (HTR-10) in China. For these reasons IG-110 is proven graphite for the HTGR application and is one of the major candidates for the Generation IV VHTR. Typical characteristics of IG-110 are given in Table 1 (Ishihara, 1991).

The properties of graphite during reactor operation are changed as a function of irradiation temperature and neutron dose. Many researchers have measured the property changes of graphite and proposed various mechanisms and relationships for these changes (*e.g.*, Simmons, 1965, Kelly, 1969). Since in-core graphite components in the VHTR will be exposed to higher neutron dose than that in the HTTR, it is important to evaluate the irradiation-induced property changes under higher neutron doses, and to develop models which can predict the irradiation-induced property changes. The property changes of irradiated graphite are strongly related to its microstructural change and therefore, it is important to investigate the graphite microstructure before and after irradiation. In order to develop a microstructural / property change model for IG-110 graphite at higher neutron doses, the relationship between changes in the bulk properties and the microstructure of IG-110 graphite have been investigated.

The method of X-ray tomography has proved promising in the investigation of the three-dimension internal microstructure of materials. This method has not only been used for non-destructive inspection in the aviation field, but also for stress analysis of porous ceramics (Aoki, 2003, Nagano, 2004). This method has also been used to investigate the three-dimension microstructure of nuclear graphite, Berre *et. al.* (2006) developed finite element models based on three-dimensional images obtained from X-ray tomography to study the relationship between the microstructure and the bulk mechanical properties of Gilsocarbon graphite (Berre *et al.*, 2006). Babout *et al.* (2008) measured detectable porosity of thermally oxidized PGA graphite directly using X-ray images and developed a thermal conductivity model (Babout *et al.*, 2008). Until recently it has been difficult to apply X-ray tomography to fine grain graphite, such as IG-110, due to the limit of resolution of the apparatus. However, recent development of image processing technologies and computer science make it now possible to apply X-ray tomography to fine grain graphite. This paper shows the applicability of X-ray tomography to the fine grain IG-110 graphite. The relationship between microstructure of IG-110 graphite and bulk electrical conductivity is also discussed.

TABLE 1: Typical properties of IG-110 graphite.

Bulk-density (g/cm^3)	Young's modulus (GPa)	Tensile strength (MPa)	Compressive strength (MPa)	Thermal conductivity (W/m · K)
1.78	7.9	25.3	76.8	80(at 400°C)

EXPERIMENTAL

Microstructural Change of IG-110 Graphite

To study the effect of microstructural change in properties of IG-110 graphite, porosity ratio in IG-110 graphite was changed by thermal oxidation. Porosity ξ can be described as follows;

$$\xi = \frac{\rho_{th} - \rho}{\rho_{th}} \qquad [1]$$

where, ρ_{th} (g/cm^3) and ρ (g/cm^3) are theoretical density and bulk density, respectively. The dimension and weight of each sample were measured before and after oxidation and the bulk density calculated. In this study, theoretical density was assumed to be 2.27 g/cm^3 (Nightingale, 1962). The thermal oxidation ratio of IG-110 graphite is expressed by using the weight loss, burn-off B as follows;

$$B = \frac{W_0 - W}{W_0} \times 100 \, (\%) \tag{2}$$

where, W_0 (g) and W (g) are weight before and after oxidation, respectively. Target values of burn-off were 0.5, 1.0, 2.0, 5.0 and 10.0 %. Ten samples, 10 x 15 x 20 (mm), were prepared for each burn-off. The graphite samples were oxidized in air at 500°C to attain uniform oxidized condition.

The electrical resistivity of the thermally oxidized IG-110 graphite samples was measured. The electrical resistivity, R ($\mu\Omega \cdot$ cm), at room temperature was calculated by using the equation based on potential drop method (Japan Carbon Association, 1968) shown as follows:

$$R = \frac{V \times A}{I \times L} \tag{3}$$

where V (V) is difference of voltage between two terminal points, I (A) is current value, A (cm^2) is cross section of sample and L (cm) is distance of terminal points. The difference in voltage between two terminal points was measured four times on each sample and the average calculated.

X-ray Tomography
X-ray tomography images of the IG-110 graphite samples were obtained at Japan Fine Ceramics Centre in Japan. An X-ray beam with focal spot size of 1µm was used with 100µA at 50kV. A charged coupled device (CCD) camera with a beryllium foil detector was used to obtain two-dimensional X-ray images for 280 layers. IG-110 graphite samples were placed at a distance of 5mm from the X-ray source to obtain an 80 times magnite and a pixel size of about 0.9µm on the CCD screen. The sample size was 3 x 3 x 2.5 (mm) for non-oxidized samples and 3 x 3 x 1 (mm) for thermally oxidized samples. Since the X-ray tomography two-dimensional slice images were taken at about 280µm intervals, sample features could be distinguished if they were larger than 280µm. The samples were thermally oxidized under the conditions as given above with target burn-off for the X-ray tomography images were 0.0, 2.0, 5.0 and 10.0%.

RESULTS AND DISCUSSIONS

Microstructural Change of IG-110 Graphite
The average value of burn–off, bulk density and porosity were given in Table 2 and the relationship between burn-off and normalized electrical conductivity (R_0/R) is shown in Figure 1. With increasing burn-off, the electrical conductivity decreases gradually. If samples had not been oxidized uniformly, that is to say samples were oxidized mainly from the surface, the difference between before and after their electrical conductivity would have

been be small. However, this was not the case, therefore it was assumed that sample oxidation was reasonably uniform.

The relationship between property change and burn-off in graphite is usually fitted to the empirical exponential decay formula;

$$\frac{S}{S_0} = \exp(-nB) \tag{4}$$

where S_0 is the property before oxidation and S is the property after oxidation, respectively. B is burn-off and n is an empirical parameter experimentally determined. There is good precedence for the use of this relationship which was first proposed by Knudsen (1959) for the relationship between Young's modulus and the porosity of a porous material. For example, Sato *et al.* (1990) confirmed that the Young's modulus of IG-110 graphite followed Equation [4] and Shibata *et al.* (2008) confirmed that the velocity change of ultra sonic waves in IG-110 graphite could also be described by Equation [4]. The relationship between electrical conductivity and burn-off based on results from this experiment study, using the Equation [4], can be expressed as follows;

$$\frac{R_0}{R} = \exp(-nB) \tag{5}$$
$$n = 5.7 \times 10^{-2}$$

This relationship is plotted in Figure 1 and is shown to give good agreement compared with the experimental data. Figure 2 shows the relationship between burn-off and porosity. The porosity can be expressed as a linear function of burn-off. Therefore, electrical conductivity can be assumed to be a function of porosity.

TABLE 2: Experimental results of burn–off, bulk density and porosity for IG-110 graphite.

Burn-off (Target, %)	Bulk density (g/cm^3)	Porosity
0	1.78	0.214
0*	1.77	0.215
0.398 (0.5)	1.78	0.212
0.955 (1.0)	1.77	0.215
1.87 (2.0)	1.76	0.221
2.26(2.0)*	1.73	0.236
4.98 (5.0)	1.70	0.247
5.17 (5.0)*	1.69	0.252
9.98 (10.0)	1.63	0.280
10.2(10.0)*	1.59	0.295

* Samples for X-ray tomography

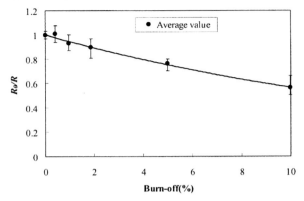

FIGURE 1: Relationship between normalized electrical conductivity and burn-off for IG-110 graphite.

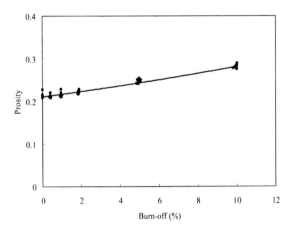

FIGURE 2: Relationship between burn-off and porosity for IG-110.

X-ray Tomography

The relationship between burn–off, bulk density and porosity for samples of IG-110 graphite examined using X-ray tomography are also given in Table 2. The Amira[8] software package was used to construct three-dimensional image from the two-dimensional X-ray images. From these images small volumes, 25×25×25 (µm), were extracted for more detailed analysis. Figures 3 show X-ray tomography images at 0% burn-off in (a) two-dimensions and (b) three-dimensions. It is easy to distinguish the graphite (filler particles and the binder) and pores visually. However, there is no visual difference between the filler particles and the binder to enable these two solid phases to be modelled explicitly.

The Simpleware[9] commercial package was used to evaluate the distribution of pores in IG-110 graphite. The ScanIP facility was used to separate pores and graphite by setting suitable thresholds to distinguish pores and graphite within the grey-scale of the raw images.

---

[8] http://www.amiravis.com
[9] http://www.simpleware.co.uk

Experimental results obtained using the mercury infiltration method along with three-dimensional density mapping functions written in Matlab were used to determine porosity thresholds. The cumulative pore volume, C_{pv} (mm³/g) can be expressed as follows:

$$C_{pv} = (\frac{1}{\rho} - \frac{1}{\rho_{th}}) \times 10^3 = \frac{\xi}{\rho} \times 10^3 \qquad\qquad [6]$$

where ρ (g/cm³), ρ_{th} (g/cm³) and ξ are bulk density, theoretical density and porosity, respectively. Since a pixel size was about 0.9μm, a pore diameter of more than 1.25μm (0.9μm×1.41) could be distinguished from the images. Experimental results obtained using the mercury infiltration method showed that the cumulative pore volume (for pore diameter greater than 1.25μm) in non-oxidized IG-110 graphite corresponded to approximately 57 (mm³/g). A theoretical cumulative pore volume was calculated using the Equation [6] to be 121 (mm³/g). The ratio of pores with a diameter of over 1.25μm was calculate to be 0.47 of the total porosity and the threshold of non-oxidized IG-110 graphite was determined as 70 in the three-dimensional-raw images. Other thresholds for 2% and 5% burn-off were determined using the same method. However, since there are no experimental mercury infiltration method results at 10% burn-off, it was assumed that the cumulative pore volume for pore diameter greater than 1.25.μm was approximately 63 (mm³/g), 1.1 times greater than for non-oxidized IG-110 graphite. Table 4 gives cumulative pore volume and threshold for IG-110 graphite samples obtained using X-ray tomography. Thus it was shown that this method can be used to evaluate the pore size and determine porosity thresholds. It is considered by the authors that this would also be a powerful method for the evaluation of pore size after irradiation.

By using the ScanFE facility, a voxel based FE mesh was created. To reduce the number of elements and computing time, the pixel size was increased to 4μm. Figure 4 shows the three-dimensional pore distribution for non-oxidized and 10% oxidized IG-110 graphite. Although some detail of microstructure of IG-110 graphite was lost due to the increase in pixel size, it can be seen from these figures that the pore distribution in IG-110 graphite can be confirmed visually and the volume and shape of the pores can be estimated from X-ray images as a function of porosity (burn-off). It suggests that three-dimensional images based on the high resolution X-ray tomography can be used to describe the microstructure of IG-110 graphite. Figure 5 shows that three-dimensional mesh models can be obtained using a relatively small number of finite elements, therefore the next stage in this work will be to export these models to the finite element software for analysis. It is considered that the effect of fast neutron irradiation on the microstructure of graphite is to close of small pores followed by the generation of new porosity due to high crystal strains. Therefore, by measuring the change in the microstructure of IG-110 using X-ray tomography, it should be possible to evaluate the influence of irradiation induced microstructural property changes in IG-110 graphite using three-dimensional finite element models. Future work will therefore involve studying the relationship between the microstructure and the bulk mechanical properties in irradiated graphite using three-dimensional finite element models.

TABLE 3: Cumulative pore volume and thresholds for IG-110 graphite samples for X-ray tomography.

Burn-off (Target, %)	Cumulative pore volume (mm³/g)	Thresholds
0.00	121	70
2.26(2.0)*	137	59
5.17 (5.0)*	149	64
10.2(10.0)*	186	64

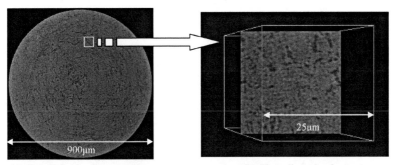

(a) Two dimensional cross section   (b) Three dimensional extract
FIGURE 3: X-ray tomography images for non-oxidized IG-110 graphite.

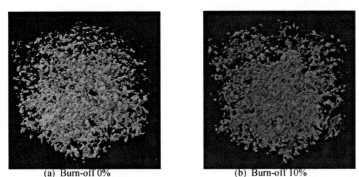

(a) Burn-off 0%   (b) Burn-off 10%
FIGURE 4: Three-dimensional pore distribution in IG-110 graphite.

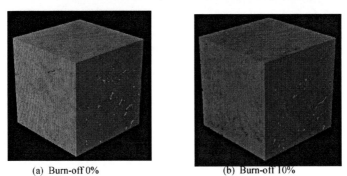

(a) Burn-off 0%   (b) Burn-off 10%
FIGURE 5: Three-dimensional finite element meshes.

CONCLUSIONS

IG-110 fine grain graphite is one of the candidate materials for core components in the Generation IV VHTR. In order to develop a microstructural / property change model for fine-grained IG-110, the relationship between changes in the bulk properties and the microstructure have been investigated. In this study, the microstructure of IG-110 graphite was modified by thermal oxidation and the change in electrical resistivity was measured. Previously, it had proved difficult to use of X-ray tomography to investigate the microstructure of fine grained graphite. However due to improvement in the resolution of the X-ray tomography equipment and the increase in computational power available in this study it has proved possible to apply X-ray tomography to the analysis of IG-110 graphite. The following results were obtained:

- IG-110 graphite samples were oxidized uniformly in air at $500°C$ and the relationship between electrical conductivity and burn-off (porosity) was obtained.

- It has been successfully demonstrated that three-dimensional images based on the high resolution X-ray tomography can be used to describe the microstructure change in IG-110 graphite due to thermal oxidation. The study has show that it should be possible to use the methodology described in this paper to predict changes in the bulk properties of irradiated IG-110 graphite.

References

Aoki, T. (2003). Micro-focus X-ray CT analysis of an orthogonal 3D-C/C composite. *Journal of the Japanese Society for Non-destructive Inspection*, 52, [10], 532-535.

Babout, L., Marsden, B. J., Mummery P. M. and Marrow, T. M. (2008). Three-dimensional characterization and thermal property modelling of thermally oxidized nuclear graphite. *Acta Materialia*, 56, 4242-4254.

Berre, C., Fok, S. L., Marsden, B. J., Babout, L., Hodgkins, A., Marrow, T. J. and Mummery, P. M. (2006). Numerical modelling of the effects of porosity changes on the mechanical properties of nuclear graphite. *Journal of Nuclear Materials*, 352, 1-5.

Ishihara, M., Iyoku, T., Toyota, J., Sato, S. and Shiozawa, S. (1991). An explication of design data of the graphite structural design code for core components of High Temperature Engineering Test Reactor. JAERI-M 91-153 (in Japanese).

Japan Carbon Association (1968). JCAS-10-1968.

Kelly, B. T. (1969). The thermal conductivity of graphite. *Chemistry and Physics of Carbons*, 5, 119-215.

Knudsen, F. P. (1959). Dependence of mechanical strength of brittle polycrystalline specimens on porosity and grain size. *Journal of the American Ceramic Society*, 42, 376-387.

Nagano, Y., Ikeda, Y., Kawamoto, H. and Takano, N. (2004). Image-based modelling by 3-D CT for homogenization stress analysis – application to porous ceramics. *Journal of the Japanese Society for Non-destructive Inspection*, 53, [11], 701-708.

Nightingale, R. E. (1962). Nuclear Graphite. *Academic Press*, New York.

Sato, S., Hirakawa, K., Kurumada, A., Kimura, S. and Yasuda, E. (1990). Degradation of fracture mechanics properties of reactor graphite due to burn-off. *Nuclear Engineering Design*, 118, 227-241.

Shibata, T., Tada, T., Sumita, J. and Sawa, K. (2008). Oxidation damage evaluation by non-destructive method for graphite components in High Temperature Gas-Cooled Reactor. *Journal of Solid Mechanics and Materials Engineering*, 2, [1],166-175.

Simmons, J. H. W. (1965). Radiation Damage in Graphite, Pergamon Press, 102-104.

U.S. DOE (2002). Nuclear Energy Research Advisory Committee and the Generation IV International Forum. 03-GA500034-08.

A Consideration of the Effect of Radiolytic Oxidation on Fracture Behaviour of Reactor Core Graphites

Andrew Hodgkins[a], T. James Marrow[b], Malcolm R. Wootton[c], Robert Moskovic[c] and Peter E. J. Flewitt[c,d],

[a]Serco TAS, Faraday Street, Birchwood Park, Warrington, WA3 66A, UK
[b]School of Materials, University of Manchester, Manchester, M1 7HS, UK
[c]Magnox North Ltd, Oldbury Naite, Oldbury-on-Severn, Bristol, BS35 1RQ UK
[d]Dept of Physics, HH Wills Laboratory, University of Bristol, Bristol, BS8 1TL, UK
Email: Andrew.Hodgkins@sercoassurance.com

Abstract

This paper provides a view on the fracture behaviour of polygranular graphites, used to moderate gas cooled nuclear reactors. Graphite is often cited as a classic example of a brittle material, because failure, in tension, is associated with small strains. However, attempts to characterise the fracture behaviour of graphite by linear elastic fracture mechanics methods have been largely unsuccessful. Observations of graphite fracture show that elastic strain energy may be dissipated by the formation of distributed micro-cracks. With continued loading, the micro-cracks open and may propagate to accommodate the strains, which can introduce non-linearity in the rising load-displacement curve. Progressive softening behaviour, after the peak load, is observed in some test specimens. This may be due to the effects of bridging in the wake of larger cracks. This type of load-displacement behaviour is characteristic of quasi-brittle materials. Radiolytic oxidation increases the proportion of porosity within reactor core graphite so that the microstructure becomes increasingly skeletal. Consideration is given to the fracture of these radiolytically oxidised graphite microstructures to support an argument for quasi-brittle behaviour.

Keywords

Reactor core graphite, Quasi-brittle, Fracture

INTRODUCTION

Graphites used for UK gas cooled reactors consist initially (the green condition) of large filler particles embedded in graphite flour and coal tar pitch binder. The reactor core bricks are fabricated either by moulding or extrusion (Gilsocarbon or Pile Grade A graphite, respectively) which after graphitising at temperatures above ~2500°C leads to differences in mechanical and physical properties of the final product (Fazluddin, 2002). Porosity remains after production at ~20% comprising open and closed pores. There are two main contributions to the changes in the microstructure and, thereby, the mechanical and physical properties of reactor core graphite when subject to the service environment. Neutron irradiation introduces defects into the crystal lattice that cluster, leading to irradiation hardening (Fazluddin, 2002). Radiolytic oxidation results in material mass loss (Best et al., 1985). Optical and high resolution electron microscopy points to a more significant increase in the volume fraction of pores within the binder region rather than the filler. The overall effect is to: (i) decrease the density of the material; (ii) increase pore size; (iii) reduce connectivity; (iv) decrease inter-pore spacing; (v) modify the distribution of the pores between filler and binder; and, (vi) reduce the overall fracture strength together with flexural strength (Kelly, 1981).

Since reactor core graphites are multi-phase, heterogeneous, polygranular materials they have been likened to other aggregate materials, such as concrete (Bazant, 1999). This link places graphite in the category of quasi-brittle materials. However the link has to be considered cautiously. This paper explores the characteristics of quasi-brittle fracture. The brittle fracture behaviour of reactor core graphites and the underlying mechanisms are considered. Attention is focussed on the change in the proportion of porosity arising from radiolytic oxidation.

QUASI-BRITTLE FRACTURE

It is appropriate before considering the fracture of reactor core graphites to address the characteristics of quasi-brittle fracture. The quasi-brittle class of materials includes concretes, mortars and porous ceramics. They all have brittle, porous, aggregate microstructures. It is the capacity of the microstructure to modify, or limit strain energy storage that produces quasi-brittle fracture behaviour.

Load-Displacement Behaviour
The classic linear elastic load-displacement curve is characterised, in tension, by a linear extension to a peak load, followed by prompt brittle fracture (Figure 1a), (Yaghi *et al.*, 2004). Non-linearity in the elastic plastic curve (Figure 1b) is associated with permanent, inelastic deformation at loads above the elastic limit. By comparison, the quasi-brittle load-displacement curve is similar to the elastic-plastic case up to the peak load (regions I & II in Figure 1c). However, the post-peak progressive softening can be observed in test specimens (Region III in Figure 1c). Post peak softening is a characteristic that attracts a great deal of attention, but it is not universally observed. Quasi-brittle materials may exhibit a wide range of different load-displacement responses.

Fracture Mechanisms
The controlling mechanisms for quasi-brittle fracture differ from those of elastic-plastic ductile materials. Non-linearity up to the peak load is associated with distributed micro-cracks (Bazant, 1999). These form in a process zone that is embedded within a stressed volume. The micro-cracks reduce the stiffness of the process zone. The post peak softening in concrete (region III) is attributed to a range of factors including friction from interlocking of aggregates (Karihaloo, 1995). However, friction bridging in the wake is governed by the roughness of the crack surfaces and shear parallel to the crack plane. Consequently, the contribution to region III from friction bridging depends on the load configuration and component or test specimen geometry. It is important to recognise that in a quasi-brittle material, the material remains locally linear elastic at all stages of the non-linear response and propagation of cracks within these heterogeneous microstructures can be progressive.

Size Effects
Quasi-brittle materials commonly show significant variability in the strength of small specimens and a change in the mean strength as the volume of the structures increases (Bazant, 1999). These 'size effects' arise from the relative size of the stressed volume and the distribution of defects, which can depend on the microstructure length-scale. Individual features in the microstructure may dominate the fracture behaviour in very small specimens. Their significance reduces as the stressed volume increases. The elastic and inelastic response of the process zone is modified, in quasi-brittle materials, by the heterogeneities in the microstructure (Figure 2). In plain specimens, the failure strain can be modified by strain relaxation from inelastic damage. With increasing process zone size, the likelihood of

inelastic damage increases, hence the probability of failure at a given applied strain decreases (Bazant, 1999). In notched specimens, the length of the inelastic zone, relative to the elastic zone is reduced by the constraints introduced due to stress intensification at the notch (L in Figure 2). The width of the damage zone is constrained by the microstructure (X in Figure 2). As a consequence, failure tends to become progressively more linear elastic as the size increases (Bazant, 1999).

By analogy to the effect of crack tip plasticity on the brittle fracture of ductile metals, non-linear behaviour and post-peak softening may not necessarily be measurable in the testing of quasi-brittle materials. For example, in notched specimens with a small process zone relative to the specimen size, the effect of inelastic damage in the process zone on the stored elastic energy in the specimen may be insignificant. A better understanding of this is needed to avoid unnecessary conservatism when relating the potential behaviour of large components to small specimen tests.

FRACTURE OF REACTOR CORE GRAPHITES

The load-displacement behaviour of unirradiated and non-oxidised nuclear graphites exhibits quasi-brittle characteristics. For example, plain specimens (*i.e.,* 25mm x 30mm cross-section) of PGA graphite, loaded in bending commonly exhibit some non-linearity prior to the peak load, followed by either prompt or progressive fracture (Figure 3(a)) (Brown *et al.,* 2006). Examples of post peak softening have been observed in notched specimens, predominantly in bending and in fracture toughness-type tests (Ouagne, 2001).

Plain specimens, of PGA graphite trepanned from Magnox reactors to monitor the effects of Radiolytic oxidation (*i.e.* weight loss), show a range of load-displacement responses and strengths for similar weight loss (Figure 3(b)), when tested in four point bending (cross-section ~1 cm^2) (Metcalfe *et al.,* 2006). In general, these show non-linearity to the peak load following an initial linear response. Occasionally, some show load drops and reduced stiffness in this range. A small number show marked post peak softening (*i.e.,* Figure 1c) (Metcalfe *et al.,* 2006). The load drops and reduced compliance indicate that damage develops before the peak load is achieved. Post test examination of the bend specimens show irregular crack paths controlled by two factors; the direction of the applied tensile stress and the heterogeneous microstructure. This is highlighted in Figure 4, which for a bend specimen of radiolytically oxidised PGA graphite (~44% weight loss). The long axis is parallel to the extension direction (Metcalfe *et al.,* 2006). The dashed line above the specimen shows the deformation following crack propagation. Fractography was undertaken using secondary electron imaging in the scanning electron microscope for un-irradiated and irradiated specimens of this axial orientation, and also perpendicular orientation. For both axial and transverse loading, relative to the brick axis, fracture links open pores within the binder regions together with some brittle transgranular cleavage fracture of filler particles (Flewitt *et al.,* 2007). The filler particles tend to show more pronounced cleavage when loading is in the traverse direction and the cracks link and propagate within the broadly aligned needle-like particles as shown schematically in (Figure 5). These observations are consistent with the differences in fracture strength between these two orientations. By comparison, fractography of irradiated bend specimens, both axial and perpendicular orientation, reveal a high density of porosity and support considerations that there is a greater proportion of porosity within the binder regions (Vatter, 2007). In general, fracture is by accumulated failure of inter-pore ligaments within the highly porous binder regions. Cracks by-pass the filler particles and follow paths within the matrix. This is highlighted in Figure 6(a), where the surface of the filler particle is evident together with adherent matrix material. By comparison in Figure 6(b) the filler particle has been completely removed. There is no evidence of cleavage in irradiated PGA graphite filler particles. Hence, there is a clear distinction between the crack propagation during fracture of un-irradiated and radiolytically oxidised high weight loss PGA graphite. Figure 6(c) and Figure 6(d) show the matrix phase of PGA graphite oxidised above 30% weight loss, in greater detail. As a result of oxidation the matrix progressively develops a pronounced skeletal or foam-like microstructure.

Damage development has been observed in-situ in Gilsocarbon graphite (Joyce *et al.*, 2008). High resolution strain surface mapping of disc compression (Brazilian disc) specimens showed a distribution of localised strained regions from the larger porosity and calcination cracks. Damage nucleated from these features, coalesced and ultimately propagated to failure (Figure 7). Sensitive strain mapping techniques have also been used to observe crack nucleation and growth on the tensile surface of Gilsocarbon flexural bend samples (Li *et al.*, 2008). Cracks with a surface length from several mm to 10s of mm developed prior to failure. These cracks were judged, from the measurement of crack mouth opening, to be shallow. Although failure often initiated from one of these defects, it was not always the longest. In a significant number of tests, the initiating defect was not observed on the sample surface prior to failure. This implies that the tensile surface of the sample developed a distribution of micro-cracks, some of which were below the resolution limit of the technique. This is the expected behaviour for a quasi-brittle material. In quasi-brittle materials containing significant stress concentrators such as a notch or macroscopic crack, micro-cracking is expected in the process zone associated with the stress concentrator. In notched specimens of unirradiated Gilsocarbon graphite, for example, it is found that micro-cracking occurs at loads above approximately 80 - 90% of the peak load (Joyce *et al.*, 2008). Distributed micro-cracks associated with propagating cracks have also been reported in a range of other grades of graphite (Sakai *et al.*, 1988).

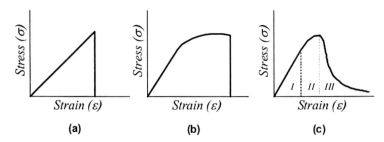

FIGURE 1: Typical stress-strain curves for (a) linear elastic materials, (b) elastic plastic materials and (c) quasi-brittle materials in tension.

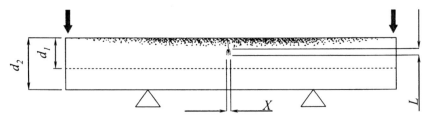

FIGURE 2: A schematic image of a plain bend geometry specimen and the effect of a sharp notch on the distribution of micro-cracks.

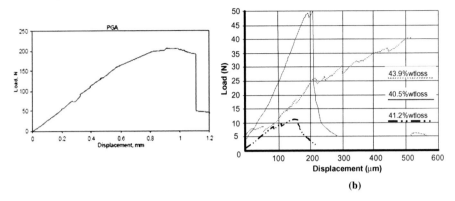

(b)

FIGURE 3: Load-displacement curve for a plain flexural specimen of (a) an un-irradiated PGA graphite and (b) irradiated PGA graphite of different wt. losses and orientations.

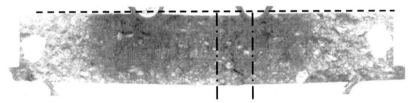

FIGURE 4: Flexural test specimen of PGA graphite, oxidised to approximately 44% wt loss. The irregular crack path is highlighted.

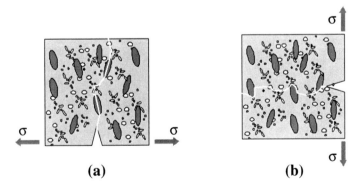

(a)   **(b)**

FIGURE 5: Schematic image showing cracking through some filler particles (a) in specimens loaded transverse to the extrusion direction and around the filler particles (b) in specimens loaded parallel to the extrusion plane.

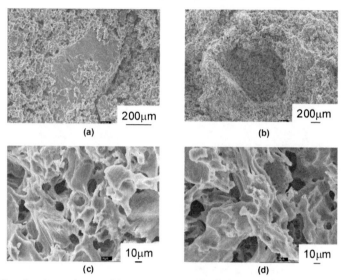

FIGURE 6: Scanning electron images of the microstructure of radiolytically oxidised PGA graphite: (a) and (b) show matrix-filler particle interfaces. The matrix is shown, at higher magnification to highlight the development of a skeletal structure in (c) and (d).

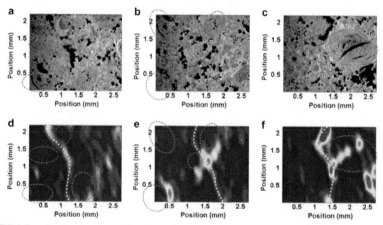

FIGURE 7: Localised strains observed on the surface of unirradiated disk specimens of Gilsocarbon graphite.

CONCLUDING REMARKS

Quasi-brittle fracture behaviour defines a class of brittle materials in which the heterogeneous and often porous microstructures provide a non-negligible contribution to the fracture process. The class is distinguished by the potential for measurable non-linear load-displacement behaviour in tension and bending, up to the peak load. The behaviour is a result of distributed micro-cracks in a damage zone. As a consequence one may question if quasi is the correct term to describe this type of fracture as pseudo, meaning "having a close resemblance to", may be more appropriate.

Polygranular graphites have been shown to exhibit many of the features associated with quasi-brittle fracture; distributed micro-cracking, non-linear load-displacement behaviour, systematic variation in strength with specimen size. Post peak softening is a distinctive feature that has been reported in graphite. However, it is not consistently observed in graphite or in other quasi-brittle materials such as concrete. This behaviour is not fundamental to a definition of quasi-brittle fracture, and can result from crack bridging processes. Rather it is the non-linear response leading to permanent deformation that characterises the materials. These graphites when exposed to service in the core of gas cooled reactors change microstructure due to the loss of mass by radiolytic oxidation. This produces an increasingly porous material. As the portion of porosity increases the microstructure develops as a skeletal framework, with an associated loss of strength. The presence of significant micro-cracking prior to the formation of macro-cracks reduces the capacity of the microstructure to store elastic energy and there is an associated relaxation of stresses. Moreover, as the mass of graphite decreases with radiolytic oxidation the propensity to form distributed micro-cracks increases prior to the formation of a dominant macro-crack.

Quasi-brittle fracture concepts provide a framework to understand the fracture behaviour of polygranular graphite test specimens, and can provide a route to prediction of the behaviour of larger components. For example, the effects of inelastic damage on elastic strain energy will influence crack propagation. In general, it may be concluded that polygranular graphites used for the construction of reactor cores have characteristics of quasi-brittle fracture commensurate with other aggregate containing materials, such as concrete.

Acknowlegement
This paper is published with the permission of Magnox North Ltd. AH and TJM are grateful to Magnox North Ltd and to British Energy Ltd for their support of the work on Gilsocarbon graphite.

References
Bazant, Z. P. (1999). Size effect on structural strength: a review. *Archive of Applied Mechanics*, 69, [9-10], 703-725.
Best, J. V., Stephen, W. J. and Wickham, A. J. (1985). *Progress in Nuclear Energy*, 16, 127-178.
Brown, M., Payne, J. F. B., Sowerbutts, I. J, Lingham, T. and Bannister, G. (2006). Investigation into graphite material variability. British Nuclear Group Report RS/E&TS/GEN/REP/0026/06.
Fazluddin, S. (2002). Crack growth resistance in nuclear graphite. Ph.D. Thesis, University of Leeds, UK.
Flewitt, P. E. J., Moskovic, R. and Wootton, M. R. (2007). A consideration of the effect of radiolytic oxidation on the directionality of fracture strength of PGA graphite. Magnox Electric North Report MEN/ESTD/GEN/EAN/0069/07.
Joyce, M. R., Marrow, T. J., Mummery, P., Marsden, B. J. (2008). Observation of microstructure deformation and damage in nuclear graphite. *Engineering Fracture Mechanics*, 75, [12], 3633 - 3645.
Karihaloo, B. L. (1995). Fracture mechanics and structural concrete. Longmans Scientific and Technical, Harlow.
Kelly B. T. (1981). Physics of graphite. Applied Science (London).
Li, H., Duff, J., Marrow, T. J. (2008). In-situ observation of crack nucleation in nuclear graphite by digital image correlation. Proceedings of PVP2008, PVP2008-61136.
Metcalfe, M. P., Brewer, N., Clucas, G. and McGrady, T. (2006). Oldbury Reactor 2 2005 installed sets - phase 3 measurements part 7 further data on strength (flexural and compressive) and dynamic Young's modulus – measurements. British Nuclear Group Report RS/E&TS/OLA/REP/0093/06.
Ouagne, P. (2001). Fracture property changes with oxidation and irradiation in nuclear graphite. Ph.D. Thesis, Bath University, UK.
Sakai, M., Yoshimura, J. I., Goto, Y. and Inagaki, M. (1988). R-Curve behaviour of a polycrystalline graphite: microcracking and grain bridging in the wake region. *Journal of the American Ceramic Society*, 71, [8], 609-616.
Vatter, I. A. (2007). Oldbury Reactor 2 2005 installed sets - phase 3', measurements part 3 flexural tests – fractography. British Nuclear Group Report MEN/ESTD/OLA/REP/0058/07.
Yaghi, A. H., Hyde, T. H., Becker, A. A. and Walker, G, (2004). The integrity of graphite blocks in nuclear reactors. University of Nottingham Report, GRA/AHY/040806.

Stress Concentration Factors for Magnox Moderator Graphite

Martin Lamb

Serco Assurance, Rutherford House, Olympus Park, Quedgeley, Glos GL2 4NF

Email: martin.lamb@sercoassurance.com

Abstract

The objective of this study is to predict stress concentration factors (SCFs) for Magnox moderator graphite. The approach is to relate theoretical SCF and measured (or effective) SCF. The values of SCF are for flexural and tensile strength tests on commercial polycrystalline graphites over the strength range 12 MPa to 94 MPa. Most of the tests are on the materials in the unirradiated condition, but some are on irradiated ex-reactor graphite. For all the tests, the effective SCFs (K_{eff}) are less than the theoretical SCFs (K_t). Notch sensitivity is used to relate theoretical and effective SCFs. Tensile test data and flexural test data are examined separately to assess whether the factors that can influence the notch sensitivity of other materials significantly affect the effective SCFs of the graphites. The factors are strength (flexural or tensile), theoretical SCF and notch radius. For both test types, notch sensitivity is not significantly affected by any of these factors. Moreover, a statistical test (Student-t) indicates that the mean values of notch sensitivity for the two test types are not significantly different. Therefore, the datasets can be combined resulting in a mean notch sensitivity of 0.28 with a standard error on the mean of ±0.034. The fracture surfaces of flexural test specimens have been compared for moderator graphite in the virgin and radiolytically oxidized conditions and these observations are used to assess the extent to which the notch sensitivity of virgin graphite can be applied to irradiated graphite.

Keywords

Stress concentration factor, Notch sensitivity, Fractography

INTRODUCTION

The theoretical stress concentration factor (SCF) is the ratio of the stress at the root of a notch to the nominal stress and can be evaluated for a specific geometry (Peterson, 1974). Materials where stresses can be redistributed by plastic deformation or the action of internal stress concentrating features (porosity or a weak second phase such as flake graphite in cast irons) are not as sensitive to notches as implied by the theoretical SCF. For these materials, the effective SCF (K_{eff}), defined as the ratio of the failure stress for an unnotched specimen to the failure stress calculated for the net section of a notched specimen without any further allowance for stress concentration, is less than the theoretical SCF (K_t).

Notch sensitivity (q) can be defined as:

$$q = \frac{K_{eff} - 1}{K_t - 1} \qquad [1]$$

so that a value of q of 0 indicates a material with no notch sensitivity and a value of 1 indicates a fully notch sensitive material.

The mechanical test data used to estimate the effect of stress concentrations on flexural and tensile strength are for commercial graphites over a range of strength from 12.0 MPa to 94.0MPa. The tests are mostly on virgin graphites, but some are on irradiated material. For these specimens, the irradiation dose is approximately 10^{23} neutrons·m^{-2} and the irradiation temperature 50°C. In addition, fractographic information is available for flexural tests on unnotched radiolytically oxidized PGA graphite.

The approach of this study has been to use the data to estimate the notch sensitivity of graphite and to use the fractographic information and mechanistic considerations to reinforce the advice for radiolytically oxidized graphites.

MODELLING

The data for a range of commercial graphites are listed in Table 1 (flexural strength) and Table 2 (tensile strength).

TABLE 1: Summary of data from notched graphite beams tested in four point bending.

graphite type	strength /MPa		SCF		radius / mm	number of observations	notch sensitivity
	plain	notched	theoretical	measured			
PGA	12.0	10.0	2.61	1.20	1.00	6	0.12
		10.4	1.88	1.15	2.40	6	0.17
	14.9	11.2	2.00	1.34	5.00	8	0.34
EK78	49.0	27.0	2.51	1.81	1.10	6	0.54
		35.0	1.86	1.40	2.50	6	0.47
Gilsocarbon	41.0	25.9	2.61	1.58	1.00	6	0.36
		30.0	1.90	1.37	2.30	6	0.41
Isotropic NI	36.1	26.9	2.90	1.34	0.64	6	0.18
		28.3	2.90	1.28	0.64	6	0.15
		24.8	3.10	1.46	0.64	6	0.22
		33.2	1.70	1.09	2.54	6	0.12
		34.3	1.70	1.05	2.54	6	0.07
		30.1	1.75	1.20	2.54	1	0.27
		28.5	1.80	1.27	2.54	6	0.33
		26.8	1.90	1.35	2.54	4	0.39
		35.6	1.20	1.01	10.16	6	0.07
	37.2	36.2	1.70	1.03	2.54	6	0.04
		32.8	1.75	1.14	2.54	6	0.18
		30.9	1.80	1.21	2.54	6	0.26
		30.3	1.90	1.23	2.54	6	0.25
Pechiney	27.0	17.0	2.51	1.59	1.10	6	0.39
		20.0	1.84	1.35	2.60	6	0.42
Poco	94.0	53.0	2.61	1.77	1.00	6	0.48
		66.0	1.90	1.42	2.30	5	0.47

TABLE 2: Summary of data from notched graphite beams tested in tension.

graphite type	strength / MPa		SCF		radius / mm	number of observations	notch sensitivity
	plain	notched	theoretical	measured			
Isotropic NA	17.0	12.5	2.90	1.36	0.64	3	0.19
		14.3	1.60	1.19	2.54	4	0.31
Isotropic NI	23.1	16.1	2.90	1.44	0.64	4	0.23
		22.5	1.60	1.02	2.54	4	0.04
Isotropic NI irradiated	47.2	33.6	2.90	1.41	0.64	3	0.21
		38.5	1.60	1.23	2.54	3	0.38
Isotropic NK	23.9	19.1	2.10	1.25	0.64	6	0.23
		11.9	3.70	2.01	0.64	5	0.37
		20.5	1.40	1.16	2.54	7	0.41
		22.5	1.40	1.06	2.54	7	0.15
		13.2	2.30	1.81	2.54	5	0.62
		12.0	2.30	1.99	2.54	6	0.76
		13.0	2.60	1.85	2.54	5	0.53
		24.8	1.10	0.96	10.16	7	-0.36
		17.8	1.50	1.34	10.16	4	0.69

Most of the estimates of effective SCF are the averages of a number of tests. For one of the graphites (isotropic NI) data are available for the virgin and irradiated conditions. The theoretical SCFs are in the range from 1.1 to 3.7. The data are presented, as a plot of effective SCF versus theoretical SCF, in Figure 1. As expected, all the effective SCFs are less than the theoretical SCFs and there is a correlation between effective SCF and theoretical SCF. On Figure 1, the various groups of data show similar paired values of theoretical and effective SCF.

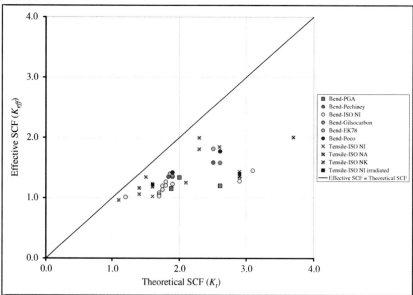

FIGURE 1: Effective SCF versus theoretical SCF for the flexural and tensile tests.

In most engineering materials, notch sensitivity is affected by strength, notch size and theoretical SCF (Peterson, 1974). To assess whether each of properties affected the notch sensitivity of the graphites, notch sensitivity is plotted against strength (flexural or tensile) (Figure 2a), theoretical SCF (Figure 2b) and notch radius (Figure 2c), respectively.

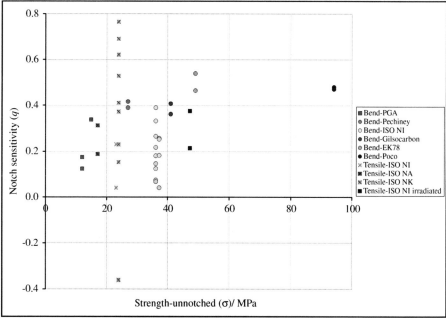

FIGURE 2a: Notch sensitivity versus strength.

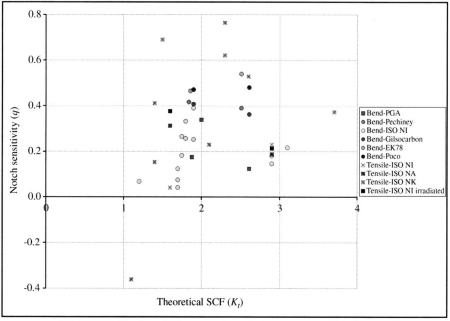

FIGURE 2b: Notch sensitivity versus theoretical stress concentration factor.

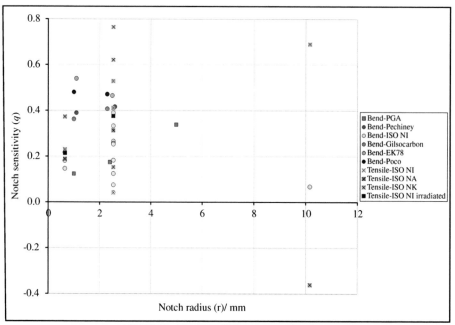

FIGURE 2c: Notch sensitivity versus notch radius.

No clear dependency is evident in any of these plots (Figures 2a, 2b and 2c). The significance of the predictors of notch sensitivity can be evaluated quantitatively by regression modelling. A suitable model is:

$$q_i = \beta_0 + \beta_1 K_{ti} + \beta_2 \sigma_i + \beta_3 r_i + \varepsilon_i \qquad [2]$$

where notch sensitivity (q) is the dependent variable and the possible predictor variables are strength (σ), theoretical SCF (K_t) and notch radius (r). β_0 to β_3 are constants estimated by regression, ε is the random error and i indexes the observations (Draper and Smith, 1981). All the observations were included in the modelling with the exception of the two data for Poco graphite. The flexural strength of this material is nearly twice that for any of the other graphites so that the pattern that emerges in a plot of notch sensitivity versus flexural strength is of a cloud of data at relatively low values of strength and two remote points for the Poco graphite (Figure 2a). Thus, the data for Poco graphite will have a strong influence on the resultant model and, since Magnox moderator graphite (PGA graphite) is a relatively low strength material, it was decided to exclude the Poco data from the analysis. The regression analysis confirmed the visual assessment of the data; none of the terms was a significant predictor of notch sensitivity at the 95 % probability level either for the flexural or tensile data taken separately or for the pooled data. Thus, the appropriate model is simply a mean value of notch sensitivity q. The mean values of notch sensitivity are:

- 0.26 with a standard deviation of ±0.140 and a standard error on the mean of ±0.030 for the flexural strength data (22 observations) and
- 0.32 with a standard deviation of ±0.280 and a standard error on the mean of ±0.072 for the tensile strength data (15 observations).

A Student-*t* test on the flexural and tensile data showed that the difference between the two means was not significant at the 5 % level. Hence the two datasets can be combined. This results in an average mean value of notch sensitivity (q) of 0.28 with a standard error on the mean of ±0.034.

FRACTOGRAPHY

Fractographic images are available from compressive and bending tests on virgin and radiolytically oxidized PGA graphite. However, to avoid complications caused by sliding of rough contacting surfaces, the images in this section are all for bending tests. PGA graphite is anisotropic and for these tests, the specimens (for graphite in the virgin and oxidized conditions) were oriented with the long edges parallel to the extrusion direction. Representative image of a specimen of virgin PGA graphite are presented in Figure 3 which shows the textured surface produced by fracture of the binder material and cleavage across filler particles over a range of magnifications.

500 µm	200 µm	100 µm
FIGURE 3a	**FIGURE 3b**	**FIGURE 3c**

FIGURE 3: SEM images of a fracture surface of a flexural test specimen of virgin PGA graphite.

Fractographic examinations have also been performed on flexural test specimens of radiolytically oxidized graphite from material used to monitor the condition of the Oldbury reactors. Three representative images from flexural test specimens are shown in Figure 4. Figures 4a and 4b are images of the fracture surface of a beam with a weight loss of 34 %. At the higher magnification, Figure 4b shows local variation in porosity which may reflect the different oxidation behaviour of the binder and the filler particles. Figure 4c shows a region of binder in a specimen with a weight loss of 44 % and more extensive porosity.

On comparison of these images, the main difference that emerges between the fracture surfaces of the virgin and the radiolytically oxidized graphite samples is the greater porosity of the latter.

APPLICATION TO RADIOLYTICALLY OXIDIZED GRAPHITE

Most of the data used to estimate the notch sensitivity of graphite were from material in the virgin condition. However the graphite moderator components of concern are subject to radiolytic oxidation. The observation that the irradiated graphite shows effective SCFs that are not distinct from the data for virgin graphites provides an indication that the notch sensitivity of irradiated graphite can be inferred from data for virgin graphite. Moreover, the main difference between the fracture surfaces of the virgin and the radiolytically oxidized graphite samples is the greater porosity of the latter and such pores tend to reduce the notch sensitivity of materials. Given the behaviour of the available data for irradiated graphite and the evidence from the fractography, this application of the data is judged acceptable.

| 500 µm | 200 µm | 200 µm |
| **FIGURE 4a** | **FIGURE 4b** | **FIGURE 4c** |

FIGURE 4: SEM image of fracture surface of flexural test specimens of radiolytically oxidized PGA graphite

CONCLUSION

For Magnox moderator graphite in the virgin and radiolytically oxidized conditions, a value of effective notch sensitivity (q) defined as

$$q = \frac{K_{\mathit{eff}} - 1}{K_t - 1}$$

of 0.28 with a standard error on the mean of ±0.034 can be used to estimate the effect of stress concentrating features. In this equation, K_t denotes theoretical SCF and K_{eff} effective SCF defined as the ratio of the failure stress for an unnotched specimen to the failure stress calculated for the net section of a notched specimen without any further allowance for stress concentration.

Acknowledgements
The author thanks the EWST Engineering Director, Magnox North for permission to present the paper and Malcolm Wootton, EWST Magnox North for constructive advice.

References
Peterson, R. E. (1974). Stress Concentration Factors. Wiley.
Draper, N. R. and Smith, H. (1981). Applied Regression Analysis, Wiley.

Using First Principles Calculations to Estimate Thermal Properties of Graphite and its Defects

Gemma Haffenden[10] and Malcolm Heggie

Dept. of Chemistry, University of Sussex, Falmer, Brighton, UK. BN1 9QJ
Email: glh20@sussex.ac.uk.ac.uk

Abstract

In order to establish the suitability of applying *ab initio* techniques to thermodynamical problems, particularly in the difficult case of graphite, some of the thermal properties of pure graphite have been calculated here, and compared to experimentally determined values. Density functional theory was used to calculate dynamical matrices and in turn lattice vibrations, and statistical thermodynamics were used to calculate total entropies, heat capacities and coefficients of thermal expansion.

Keywords

Graphite, Heat capacity, Thermal expansion

INTRODUCTION

The thermodynamical properties of graphite are of great interest to material scientists, due to its highly anisotropic nature and its negative thermal expansion in the basal plane (*a* axis) at lower temperatures. As an example of a layered material, it serves theoreticians well, although its uses as a nuclear moderating material and its relation to graphene and nanotubes mean that in depth knowledge of the thermal properties of graphite and graphitic materials is very important.

Previous theoretical investigations into the thermal properties of graphite, such as Mounet and Marzari (2005), have required the use of experimentally determined parameters, such as the *c/a* ratio, in order to reach agreement with experimental values. In this work, we obtained good agreement with experimental values without the use of any fitting or experimental parameters. Density functional theory (DFT) could in principle be exact, but our ignorance of the true exchange-correlation functional requires us to adopt approximations to it and one of the simplest is the local density approximation (LDA). The validity of the LDA for layered materials, and graphite in particular, has often been called in to question (*e.g.* Kohn *et al.* 1998), although recent work reaffirms its use for graphite interlayer properties. While LDA is known not to reproduce dispersion forces in rare gases correctly, Girifalco and Hodak (2002) show that, if the details of the calculation are sufficiently accurate, graphite interlayer interactions are well described if the separation is not more than 15% away from the equilibrium separation. Long range dispersion forces must be included at larger strains. The generalized gradient approximation (GGA) was introduced as a refinement of the LDA (Perdew *et al.* 1986). Instead of the functional depending only on the charge density at a given point, it included a dependence on the gradient of density. Ooi *et al.* (2006) showed that LDA does generate some interplanar bonding and bulk properties similar to experimental values, but that GGA fails to produce interplanar bonding. This is in spite of the fact that

---

[10] Author to whom any correspondence should be addressed.

neither approximation reproduces dynamical Van der Waals bonding, and their calculated electronic structures are effectively identical.

We have used LDA only, with no scaling or experimental contributions, to calculate thermal properties such as heat capacity and coefficient of thermal expansion. We see a good agreement between our calculated heat capacity dependence on temperature and experimental values. We also see excellent agreement between our *a*-axis coefficient of thermal expansion dependence on temperature and experimental results. The *c* axis coefficient of thermal expansion at higher temperatures proves sensitive to the details of the calculation although the low temperature results are promising.

METHOD

We have performed a DFT study on 36 and 64 atom, hexagonal, AB stacked graphite unit cells. The AIMPRO supercell code was used, fitting the charge density to plane waves with an energy cut-off of 200 Ry. The exchange correlation energy has been determined using the LDA, with the functional as parameterised by Perdew and Wang (1992). Gaussian basis functions centred at atom sites are used to construct the many electron wave function. Each radial Gaussian is multiplied by angular momentum expansions up to $l = 0$ (labelled s), $l = 1$ (labelled p) or $l = 2$ (labelled d). Orbital functions; using this nomenclature the basis set pdpp was used for all of the following calculations. Larger basis sets of up to pdddp were tested, although pdpp was found to be sufficiently accurate for our purposes. Norm-conserving pseudo potentials were used, these being based on the Hartwigsen-Goedecker-Hutter scheme (1998). A Brillouin zone sampling grid using the Monkhorst-Pack (1976) special *k*-point technique was chosen to suit each unit cell. The density of *k*-points was kept constant as much as possible, and an increase in density of *k*-points was tested to ensure good convergence of total energies. The oscillation of charge density in part-filled degenerate orbitals during the self consistency cycle was damped using a Fermi occupation function of kT=0.04 eV.

The vibrational modes of a structure can be calculated from the energetic double derivatives of the fully optimised structure, with respect to displacement of each of the atoms. Each atom is displaced a distance, δ (usually 0.02 a.u.) along each axis, then the charge density is recalculated giving the forces, $f^{\pm}_{mb}$, on each atom in each direction. The double derivatives of the energy with respect to the displacement of atoms *a* and *b* is approximated by

$$D_{la,mb} = \frac{f^{+}_{mb}(l,a) - f^{-}_{mb}(l,a)}{2\varepsilon} \qquad [1]$$

where *l* is the displacement axis, and *m* labels the direction of the force produced. These double derivatives divided by the square roots of the atomic masses of *a* and *b* give the force constants of the system in the form of the dynamical matrix. The double derivatives are calculated using AIMPRO; the calculation of the dynamical matrix and its subsequent diagonalisation is performed by another code, DIAGDD. Due to the step size used, this method incorporates some anharmonic contribution; the resulting vibrations are hence described as 'quasi-harmonic' (Jones *et al.* 1994).

In short, all vibrations are given by diagonalisation of the matrix of second derivatives and heat capacities and coefficients of thermal expansions are given from the vibrations by use of statistical thermodynamics as described below.

FIRST PRINCIPLES CALCULATION OF HEAT CAPACITY

The molar heat capacity of a substance is defined as the amount of energy required to raise the temperature of one mole of a substance by one degree (K), which at constant volume is:

$$C_V = \left(\frac{\partial E}{\partial T}\right)_V \qquad [2]$$

Experimental studies have shown that the heat capacity of graphite exhibits T^2 dependence at low temperatures, as opposed to the expected T^3 behaviour (Krumhansl *et al.* 1953). Both the Einstein model and Debye model have been used in this work in order to establish the suitability of each to the study of graphite. The main assumption of the Einstein model is that all atoms vibrate with the same frequency, which gives the Einstein temperature of the material and therefore the specific heat (Donovan 1971):

$$C_V = 3RG\left(\frac{\Theta_E}{T}\right), \qquad \Theta_E = \frac{\hbar\omega_E}{k_B} \qquad [3]$$

where G is a function of the Einstein temperature Θ_E and temperature T. The main drawback of this Einstein model is the assumption that all atoms vibrate with one frequency, not the range of frequencies we know from the phonon spectrum. We have used the sum over modal frequencies in a modified Einstein model (Barrera *et al.* 1995) as below:

$$C_V = \sum_q c_{V,q} = k_B \sum_q \left(\frac{\hbar\omega_q}{2k_BT}\right)^2 \frac{1}{\sinh^2\left(\frac{\hbar\omega_q}{2k_BT}\right)} \qquad [4]$$

If $c_{V,q}$ is calculated individually for each mode of vibration of the material, the sum of these results gives a much more complete picture of the total heat capacity. At low temperatures the experimental heat capacity for most materials follows a T^3 relation; the Einstein model however does not follow this relation. The low temperature behaviour (for most materials) has been improved upon with the Debye model. The Einstein model is often used as a high temperature approximation, in cases where optical modes dominate.

The Debye model was developed for use realistically only with acoustic modes of vibration (Donovan 1971). The Debye specific heat is defined as:

$$C_V = 9Nk_B\left(\frac{T}{\Theta_D}\right)^3 \int_0^{\Theta_D/T} \frac{x^4 e^x}{\left(e^x - 1\right)^2} dx, \qquad \Theta_D = \frac{\hbar\omega_D}{k_B} \qquad [5]$$

The Debye temperature Θ_D, is produced from the Debye frequency ω_D, which is the highest frequency mode present in the material.

Both models were used in this work and although the Debye model was found to be marginally closer to experiment at high temperatures, the modified Einstein model was better over the whole range and far more appropriate for low temperatures. Literature values for irradiated graphites are also shown.

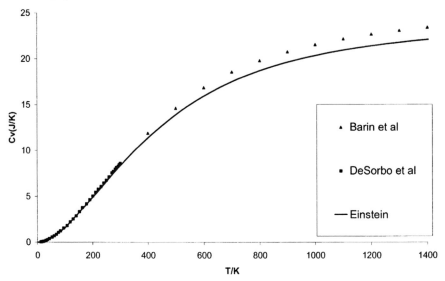

FIGURE 1: Heat Capacity of graphite as a function of temperature up to 1400 K. A comparison of the Modified Einstein model (present work) to experimental values from Barin *et al* (1989), DeSorbo *et al* (1953) and Desorbo *et al* (1957). The DeSorbo results are for both virgin and irradiated graphite – see blow up of low T region in Figure 2.

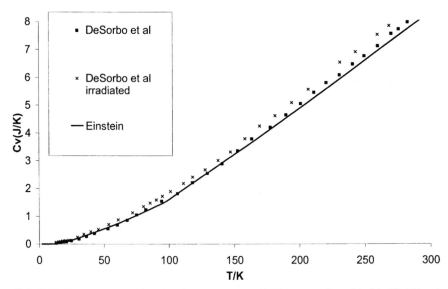

FIGURE 2: Heat Capacity as a function of temperature up to 300 K. A comparison of the Modified Einstein model (present work) to experimental values at low temperatures for virgin and irradiated graphite from. DeSorbo *et al.* (1953) and DeSorbo *et al.* (1957).

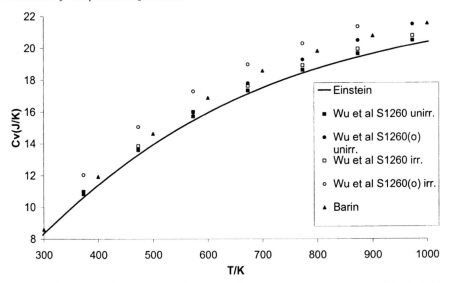

FIGURE 3: Heat Capacity as a function of temperature from 300 to 1000 K. A comparison of the Modified Einstein model (present work) to experimental values for virgin graphite from Barin *et al.* (1989) and for virgin and irradiated graphite Wu *et al.* (1994).

The results of DeSorbo *et al.* (shown in Figure 2) and those of Wu *et al.* (shown in Figure 3) show an increase in the heat capacity of the graphite after irradiation. In general, the modified Einstein model gives the best account of experimental results.

The difference between the results of the modified Einstein model and the experimental values increases with temperature, this may be due to the electronic contribution to the heat capacity, but is most likely due to the omission of structural changes associated with thermal expansion of the graphite at higher temperatures.

FIRST PRINCIPLES CALCULATION OF THE COEFFICIENTS OF THERMAL EXPANSION

In general vibrational frequencies tend to decrease as volume increases, and *vice versa*. This relation between the vibrational modes and volume of the material is expressed through the mode Gruneisen parameter (Dove, 1993):

$$\gamma_q = -\left(\frac{\partial \ln \omega_q}{\partial \ln V}\right), \qquad \gamma = \frac{\sum c_{V,q}\gamma_q}{\sum c_{V,q}} \qquad [6]$$

Graphite is a highly anisotropic material and as such has very different thermal expansion properties perpendicular and parallel to the basal plane; so two coefficients of thermal expansion are required to describe the overall thermal behaviour of the material. The elastic compliances, heat capacity and Gruneisen parameters of the material are used to calculate the coefficient of thermal expansion in the two different directions.

$$\alpha_\perp = \frac{c_V}{V}\{(s_{11} + s_{12})\gamma_\perp + s_{13}\gamma_=\}\qquad\qquad\qquad [7]$$

$$\alpha_= = \frac{c_V}{V}\{2s_{13}\gamma_\perp + s_{33}\gamma_=\}\qquad\qquad\qquad [8]$$

where c_V is the heat capacity, V is the molar volume of graphite, s values are elastic compliances and $\gamma_\perp$ and $\gamma_=$ are the Gruneisen parameters perpendicular to and parallel to the basal plane, respectively. All but the compliances were calculated using *ab initio* techniques. The molar heat capacity and modal heat capacities were as calculated from the first section and the compliances used were the latest 'best fit to experiment' taken from Cousins and Heggie (2003). The mode Gruneisen parameters were calculated from the gradient of a linear fit to phonon frequencies at a series of expansions of either the *a* or *c* axis at intervals of 0.5% up to 5, 6 and 7 % maximum expansions.

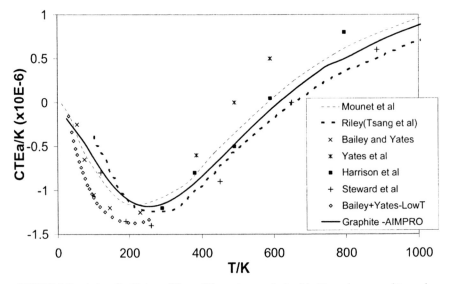

FIGURE 4: Basal plane Coefficient of Thermal Expansion as calculated in this work compared to previous theoretical work and experimental data from Tsang *et al.* (2005), Bailey and Yates (1970) and data from Kelly (1981).

The calculated coefficient of thermal expansion of the basal plane for graphite fits the experimental results for the temperature range 0 – 1000 K. The details of the calculation, such as the extent of the expansion (5, 6 or 7 % maximum) and the amplitude of the displacement of the atoms in the calculation only have a negligible effect on the final results of the *a* axis thermal expansion.

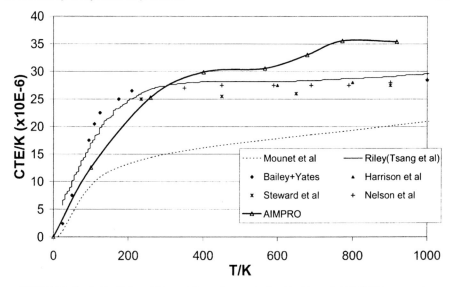

FIGURE 5: C-axis Coefficient of Thermal Expansion as calculated in this work (AIMPRO) compared to previous theoretical work and experimental data from Tsang *et al.* (2005), Bailey and Yates (1970) and data from Kelly (1981).

The calculation of the coefficient of thermal expansion of the *c* axis has proven more of a challenge for reasons such as the limitations of the LDA and the unit cell sizes. A modified method was adopted for the calculation of the *c* axis CTE. Mode Gruneisen parameters were calculated piece-wise for strains of 0.25, 0.75, 1.25, 1.75, 2.25, 2.75 and 3.25 %. Then the CTE(T) was plotted for each of these strains, and absolute strain was calculated by integration of CTE(T). Where the chosen strains and the strains obtained by integration coincided, a data point for CTE(T) was obtained (Triangles shown in Figure 5). This method is in the process of being refined, but is giving good results so far, and appears to be an improvement over Mounet and Marzari's linear response theory results (2005). The sensitivity of the c axis results may tell us more about both the material and the method upon further investigation.

CONCLUSIONS

The calculation of thermodynamic properties of pure graphite is a challenge due to the various approximations necessary to implement a workable method. This work is a contribution to a better understanding of both the material itself and an understanding of how to approach the study of its thermal properties. We have established a method of calculation of the heat capacity for a large range of temperatures from first principles. Experimentally determined values of heat capacity for irradiated graphite have also been included to show that there is an increase in the heat capacity of the graphite upon irradiation. This method of calculation can now be used to study the causes of this increase.

The calculation of the coefficient of thermal expansion in the basal plane (*a* axis) has proved successful. The calculated results compare well to experimentally determined values. The calculated coefficient of thermal expansion for the *c* axis has proved more difficult to determine accurately. Parameters such as the amount of expansion included in the calculation and the amplitude of the vibrations have a large effect on the overall results. The LDA and unit cell size are likely to be causing problems for the calculation of bulk properties along the c axis due to the treatment of the interlayer bonding and long range vibrations. Our results are promising, however, and appear to be an improvement over previous theoretical thermal expansion results.

Acknowledgements
The authors would like to thank British Energy Generation Ltd for financial support and permission to present this paper. The views expressed in this paper are those of the authors and do not necessarily represent those of the British Energy Generation Ltd.

References
Barin, I. (1989). Thermochemical Data of Pure Substances, Parts I and II. VCH.
Bailey, A. C, Yates, B. (1970). Anisotropic Thermal Expansion of Pyrolytic Graphite at Low Temperatures. *Journal of Applied Physics,* 41 [13] 5088.
Barrera, G. D, Bruno, J. A. O, Barron, T. H. K, Allan, N. L. (1995). Negative thermal expansion. *Journal of Physics-Condensed Matter* 17 [4] R217-R252
Cousins, C. S. G, Heggie, M. I. (2003). Elasticity of carbon allotropes. III. Hexagonal graphite: Review of data, previous calculations, and a fit to a modified anharmonic Keating model. *Physical Review B* 67 [2] 024109.
Desorbo, W, Tyler, W. W. (1953). The Specific Heat of Graphite from 13-Degrees-K to 300-Degrees-K. *Journal of Chemical Physics* 21 [10] 1660-1663.
Desorbo, W, Tyler, W. W. (1957). Effect of Irradiation on the Low-Temperature Specific Heat of Graphite. *Journal of Chemical Physics* 26 [2] 244-247.
Donovan, B. Angress, J. F. (1971). Lattice Vibrations. Chapman and Hall.
Dove, M. T, (1993). Introduction to Lattice Dynamics. Cambridge University Press.
Girifalco, L. A, Hodak, M. (2002). Van der Waals binding energies in graphitic structures. *Physical Review B.* 65 [12] .
Hartwigsen, C, Goedecker, S, Hutter, J. (1998). Relativistic separable dual-space Gaussian pseudopotentials from H to Rn. *Physical Review B* 58 [7] 3641-3662.
Jones, R, Goss, J, Ewels, C, Oberg, S. (1994). Ab-Initio Calculations of Anharmonicity of the C-H Stretch Mode in HCN and GaAs. *Physical Review B* 50 [12] 8378-8388.
Kellett, E. A, Jackets, B. P, Richards, B. P (1964). A Study of the Amplitude of Vibration of Carbon Atoms in the Graphite Structure. *Carbon* 2 [2] 175-183.
Kelly, B. T. (1981). Physics of Graphite. Applied Science Publishers Ltd, Barking, UK. ISBN 0-85334-960-6.
Kohn, W, Meir, Y, Makarov, D. E. (1998). VanderWaals energies in density functional theory. *Physical Review Letters* 80 [19] 4153-4156.
Krumhansl, J, Brooks, H. (1953). The Lattice Vibration Specific Heat of Graphite. *Journal of Chemical Physics* 21 [10] 1663-1669.
Monkhorst H. J, Pack J. D. (1976). Special Points for Brillouin-zone integration. *Physical Review B* 13 5188.
Mounet, N, Marzari, N. (2005). First-principles determination of the structural, vibrational and thermodynamic properties of diamond, graphite, and derivatives. *Physical Review B* 71 [21].
Ooi, N, Rairkar, A, Adams, J. B. (2006). Density functional study of graphite bulk and surface properties. *Carbon,* 44 [2] 231-242.
Perdew, J. P, Yue, W. (1986). Accurate and Simple Density Functional for the Electronic Exchange Energy - Generalized Gradient Approximation. *Physical Review B* 33 [12] 8800-8802.
Perdew, J. P, Wang, Y. (1992). Accurate and Simple Analytic Representation of the Electron-Gas Correlation-Energy. *Physical Review B* 45 [23] 13244-13249.
Tsang, D. K. L, Marsden, B. J, Fok, S. L, Hall, G. (2005). Graphite thermal expansion relationship for different temperature ranges. *Carbon* 43 [14] 2902-2906.
Wu, C. H, Bonal, J. P, Kryger, B. (1994). The effect of high-dose neutron irradiation on the properties of graphite and silicon carbide. *Journal of Nuclear Materials* 208 1-7.

Irradiation Induced Creep in Nuclear Graphite – Present Status and Near Future Activities

Gerd Haag

Nuclear Technology Consultancy, Am Malefinkbach 3, D-52441 Linnich, Germany
Email: gdhaag@t-online.de

Abstract

Irradiation induced creep in nuclear graphite is an important problem in gas-cooled nuclear reactors for assessing the lifetime of graphite parts exposed to higher fast neutron fluences. Limitations of the available creep models and the needs of older as well as new gas-cooled reactors necessitate new efforts aiming beyond the existing knowledge. British Energy therefore organised a workshop on this issue setting as a goal to verify the present status of theoretical and experimental research, and to find out desirable and achievable next steps. NRG (Petten, NL), ORNL (Oak Ridge, USA), and INL (Idaho Falls, USA) offered to use their facilities for new irradiation creep experiments.

Keywords

Nuclear graphite, Irradiation-induced creep, Creep models

INTRODUCTION

The purpose of the recent workshop organised by British Energy was to engage the international nuclear graphite community to help in the understanding about irradiation induced creep in graphite and to identify the needs and goals for a future irradiation creep experiment. This desire originated from experiences with graphite bricks ageing in British Energy AGRs, from the imperfection of existing creep models and from the current lack of data in combination with the complexity of the matter.

Brick cracking is caused by stresses exceeding the current materials strengths. Therefore, the understanding of the processes driving stress build-up and release in graphite is essential. Only irradiation creep enables graphite to withstand irradiation-induced strains of several percent. The benefits of a better understanding of creep are cost reduction due to less core inspections, a reduced risk of brick cracking, and availability of the plant over longer lifetimes.

So, with respect to some of the problems of existing graphite reactors and in order to avoid similar problems in the next gas-cooled reactors to be built near future activities are necessary. However, they should be based upon the present status of knowledge and experience which is first summarised here.

FIRST EMPIRICAL CREEP MODELS

Research on irradiation induced creep in graphite has been going on for 50 years. During the first 20 years, two important findings resulted from numerous creep experiments: The primary (or transient) creep strain ε_I was confirmed to be nearly equal to one elastic strain σ/E_0. Its transition to higher fluences can be described by an additional exponential expression rapidly decreasing with increasing fluence γ in units of $10^{24}/m^2$ (EDN):

$$\varepsilon_1 \; = \; \frac{\sigma}{E_0} \cdot \left[1 - \exp\!\left(-4 \cdot 10^{-24} \cdot \gamma\right)\right]. \qquad\qquad\qquad [1]$$

The secondary (or steady state) creep strain ε_2 was observed at fluences $> 10^{24}/m^2$ (EDN) (Kelly 1977). When data for ε_2 are plotted against neutron fluence they are different for different materials, but when plotted as creep strain/unit initial elastic strain σ/E_0 a single curve is obtained. The numerical expression of these results is:

$$\varepsilon_2 \; = \; k \cdot \gamma \cdot \frac{\sigma}{E_0} \qquad\qquad\qquad [2]$$

The factor $k = 0.23 \cdot 10^{24}/m^2$ (EDN) was called "creep constant" and was valid up to the highest fluences realised until then, and in the temperature range 140 to 650°C. The data existing at that time were interpreted in the way that the creep strain is:

- Substantially independent of irradiation temperature over the range 300 to 600 °C;
- Independent of the mode of stressing (tensile or compressive);
- Linear with applied stress (except at high stresses);
- Uniquely expressed in terms of elastic strain units for different graphites (Brocklehurst 1976).

In their theory of irradiation creep in graphite, Kelly and Brocklehurst (1989) relate the macroscopic creep rate to the deformation of the individual crystallites. If creep is caused by basal plane slip, the creep rate in direction xx is:

$$\frac{d\varepsilon_{xx}}{d\gamma} \; = \; K \cdot \sigma_{xx} \cdot \left(\frac{c_{44}}{E_{xx}}\right) \qquad\qquad\qquad [3]$$

where σ_{xx} is the uniaxial stress in the x-direction, (macroscopic stress),
c_{44} is the single crystal elastic constant for basal shear,
E_{xx} is the Young's modulus of the polycrystal in the x-direction,
K is a constant.

At low fluences where dislocation pinning occurs, irradiation damage does not affect the creep rate since c_{44} and E_{xx} are increased in the same way. Later on, when structural effects caused either by oxidation or by dimensional changes due to irradiation the creep rate will be affected by related changes in E_{xx}.

Kelly and Brocklehurst (1977) considered the relative change in Young's modulus due to fast neutron irradiation to be the product of a factor $(E/E_0)_p$ due to dislocation pinning, and a factor $(E/E_0)_s$ due to the dimensional changes of the crystallites. Both effects occur in almost completely different fluence ranges which allow their separation. When at the plateau of the Young's modulus vs fluence function, the pinning component reaches saturation and structural changes have not yet occurred let the Young's modulus be equal to E'. Then the structural factor is:

$$S(\gamma) = \left(\frac{E(\gamma)}{E_0}\right)_s = \frac{E(\gamma)}{E'}. \tag{4}$$

According to Equation [3] only this structure factor affects the creep rate and is therefore introduced in Equations [1] and [2]:

$$\varepsilon_c = \varepsilon_1 + \varepsilon_2 = \frac{\sigma}{S(\gamma)\cdot E_0}\left[1-\exp(-4\gamma)\right] + 0.23\cdot\frac{\sigma}{S(\gamma)\cdot E_0}\cdot\gamma. \tag{5}$$

The effect of radiolytic oxidation has also been considered. If radiolytic weight loss alone reduces the Young's modulus from E_0 to E_w then $S(\gamma)\cdot E_0$ in Equation [5] is replaced by $S(\gamma)\cdot E_w$. Thus, the improved model (sometimes called the UK Model) was assumed to cope with structural changes which is a most prominent effect if irradiation damage at higher fluences beyond turn-around of the dimensional changes. Another difference to the preceding model is that the introduction of the structure factor implies dependencies upon irradiation temperature and graphite grade. It should also be mentioned that there is still no difference in the theoretical creep behaviour between tensile and compressive creep.

MODELLING CREEP AT HIGHER FLUENCES

UKAEA experiments gave creep rate data extending to about $5\cdot10^{25}/m^2$ (EDN). In order to check how far the UK creep model can be extrapolated to higher fluences, KFA Jülich have performed several constant stress creep experiments in both tension and compression. The highest fluences attained were between 25 and 28 dpa, and temperatures of 300 and 500 °C (Haag 2005).

These new data showed clearly that the secondary creep state is followed by a tertiary state where the creep rates are much smaller than predicted. From this result it is obvious that the structural term $S(\gamma)$ in Equation [5] is not strong enough to describe the stiffening of the material caused by shrinking (increasing density). Tensile as well as compressive creep coefficients exhibit turn-around behaviour similar to that of volume changes. Therefore, Kennedy *et al.* (1988) have introduced a new structural factor replacing the constant k in Equation [2] by a new parameter:

$$k' = k\cdot\left[1-\mu\cdot\frac{\left(\frac{\Delta V}{V_0}\right)}{\left(\frac{\Delta V}{V_0}\right)_{min}}\right] \tag{6}$$

where $\Delta V/V_0$ is the volume change with fluence and $(\Delta V/V_0)_{min}$ is the maximum volume change at turnaround. The value μ was found to be independent of temperature equal to 0.75.

For tensile creep at 300, 500, and 900°C the Kennedy Model is in excellent agreement with the data. However, beyond turn-around, compressive creep data differ markedly from the tensile data, and this difference cannot be accounted for by any of the structure factors that have been used so far.

The difference between creep under tensile and under compressive load is not a surprise. Under compression there is only a limited pore volume existing to compress the material. Under tension there are no limits for expanding a graphite sample. Therefore, it was much more surprising that Kelly and Brocklehurst (1977) had claimed that the creep strain would be independent of the mode of stressing (tensile or compressive) (Brocklehurst *et al.*, 1976).

In the fluence range up to $1.5 \cdot 10^{26}/\text{m}^2$ (EDN) equivalent to about 20 dpa the Kennedy model is a reasonable tool to describe irradiation induced creep. It was validated at three very different temperatures and different loads, but its limits have also become obvious indicating that there is no use to improve the model since it is predominantly an empirical model. The discrepancy between tensile and compressive creep data has shown that structural features of polycrystalline graphite need to be considered in the development of follow-up models, *e.g.*, the significance of different kinds of porosity or the CTE changes due to irradiation with and without creep stress.

A severe problem for the application of the Kennedy model to AGR graphite is caused by the higher degree of oxidation which affects the volume turn-around being the basis for the creep coefficient. Before coming to further creep models still developed it may be useful to remember the facts that currently are widely agreed:

- There is primary or transient (recoverable) creep of about 1 elastic strain unit.
- There is secondary or steady state (non-recoverable) creep which is also proportional to the applied stress.
- Primary and Secondary creep measured in initial elastic strain units ($\varepsilon_c E_0/\sigma$) are independent of the graphite grade.
- Creep strain changes the coefficient of thermal expansion (increasing in compression and decreasing in tension).

ADVANCED MODELS CURRENTLY BEING DEVELOPED

This last bullet point inspired new activities to find a formalism which is able to decide between the two strain modes. In addition a correction of the over-prediction of creep strain at high fluence is desired. A first attempt in this direction was made by Kelly and later on continued by Kelly and Burchell (1994) since it became obvious that the previous models were inadequate. Their new ideas are based on an old theory which can be traced back to the late 1950s and early 1960s when Reynolds and Simmons described the structure of polycrystalline graphite as a pile of carbon crystallites with some porosity in between. Dimensional changes resulting either from temperature changes (CTE) or from the Wigner effect both occur within the crystallites. However, they are not completely transmitted to the properties of the whole body since pores can accommodate dimensional changes to some extent.

This model suggests itself that there should be a relation between thermally and irradiation induced dimensional changes. The dimensional change rates due to irradiation can be written as:

$$\frac{dG_x}{d\gamma} = \frac{\alpha_x - \alpha_a}{\alpha_c - \alpha_a} \frac{dX_T}{d\gamma} + \frac{1}{X_a} \frac{dX_a}{d\gamma} \quad \text{where} \quad \frac{dX_T}{d\gamma} = \frac{1}{X_c} \frac{dX_c}{d\gamma} - \frac{1}{X_a} \frac{dX_a}{d\gamma} \qquad [7]$$

with α_x, the graphite thermal expansion coefficient in direction x, $\alpha_c - \alpha_a$, the difference in crystal thermal expansion coefficients, γ, the neutron fluence, and X_T being the crystal change parameter.

In case of an imposed creep strain this formula needs to be changed since thermal expansivity is affected by creep strain:

$$\frac{dG_x^1}{d\gamma} = \frac{\alpha_x^1 - \alpha_a}{\alpha_c - \alpha_a} \frac{dX_T}{d\gamma} + \frac{1}{X_a} \frac{dX_a}{d\gamma} \qquad [8]$$

Only the difference between Equations [7] and [8] should be the basis for calculating the creep strain rate according to its definition as the difference in dimensional changes between a stressed sample and an unstressed sample irradiated under identical conditions. The creep rate can therefore be written as:

$$\varepsilon_c^T = \varepsilon_c^A - \int_0^\gamma \left(\frac{\alpha_x^1 - \alpha_x}{\alpha_c - \alpha_a} \right) \left(\frac{dX_T}{d\gamma} \right) d\gamma \qquad [9]$$

For exact evaluation of the existing high fluence creep data according to Equation [9] better information is needed about the parameters involved, first of all about α_x^1. Besides others, this will be a major task for future irradiation creep experiments.

Today there is a lot of discussion about the influence of stress on CTE, and if it might indicate that there is a dimensional change process which has to be separated from the "true" creep effect. There is also some hope that the true creep strain will be (again) a linear function of fluence even at higher fluences. However, since it is currently not clear if this model will be the ideal solution it is recommended to measure as many properties as possible on the crept as well as the non-crept irradiation samples.

Simmon's (1965) theory has always been assumed to be valid only if structural changes such as the formation of new porosity do not occur. Looking for a creep model for high fluences needs additional measures to cope with this difficulty. Recent results look promising but need also verification and validation.

As a consequence of identification of a number of shortfalls in current models and of a revisit of available creep data, work by Davies and Bradford (2008) came to quite another model for irradiation induced creep, raising immediately the question how different two models can be both claiming to be applicable to the same sets of data and to be able to come to the same predictions. Nevertheless, it is always useful to tackle a difficult problem from different directions.

Bradford and Davies describe creep strain by a formula following the formulation for primary and secondary creep, but extended by expressions exhibiting semi-empirical character.

$$\varepsilon_c = \frac{\alpha k_1}{E_0} \exp^{-k_1\gamma} \int\limits_{\gamma'=0}^{\gamma} \frac{\sigma}{W} \exp^{k_1\gamma'} d\gamma' + \frac{\beta}{E_0} \int\limits_{\gamma'=0}^{\gamma} (1-\exp^{-k_2\gamma'}) \frac{\sigma}{SW} d\gamma' + \frac{\omega k_3}{E_0} \exp^{-k_3\gamma} \int\limits_{\gamma'=0}^{\gamma} \frac{\sigma}{W} \exp^{k_3\gamma'} d\gamma' \qquad [10]$$

$\underbrace{\qquad\qquad\qquad\qquad}_{\text{Primary Creep}}$ $\underbrace{\qquad\qquad\qquad\qquad}_{\text{Secondary Creep}}$ $\underbrace{\qquad\qquad\qquad\qquad}_{\text{Recoverable Creep}}$

with E_0 unirradiated static modulus,
      S Structure term
      W term representing oxidation
      α 1 elastic strain unit,
      β, ω, k_1, k_2, k_3 constants to be adjusted.

Despite encouraging first attempts to describe existing data using the Bradford/Davies model, new data from stressed and unstressed graphite irradiations are required in order to test the model. In particular, attention has to be paid to the *phenomena* recoverable and non-recoverable creep.

MAJOR SHORTFALLS OF THE EXISTING MODELS

In contrast to, *e.g.*, the UK model, appropriate models must not break down at high fluences. Therefore they must account for structural changes such as densification (mainly before turn-around) and pore generation (mainly after turn-around and due to oxidation). In contrast to, *e.g.*, the UK model or the Kennedy model, appropriate models must be able to make a distinction between the effects of compressive and tensile stresses. Current understanding is that the effect of the two different stress modes is mainly on thermal expansion coefficients. Therefore, it is assumed that studying irradiation induced changes of CTE with and without stress might give a clue to the basic structural processes of creep. In the end, it must become clear whether the two creep models Kelly/Burchell and Bradford/ Davies are either equivalent or even identical, *i.e.*, making the same predictions, or to find out which one is valid in which situation. As far as irradiation induced processes within graphite crystallites, *e.g.*, the correlation between creep rate and crystal strain rate, are the basis for some of the currently existing models this needs confirmation. Therefore, it may be necessary to include highly-oriented pyrolytic graphite (HOPG) samples in the investigations. None of the existing models so far is able to predict lateral creep strain including its fluence dependence. The question if uniaxial creep experiments will provide enough information or if biaxial testing is inevitable has not finally been answered.

DESIGN CONDITIONS FOR BRITISH ENERGY'S CREEP TESTS

Some boundary conditions for near future research activities on irradiation induced creep can be derived from the situation of British Energy's AGRs.

- The range of relevant temperatures is 350 to 500°C.
- Maximum weight losses due to radiolytic oxidation should be 40%.
- Currently accumulated fluences of AGR bricks are at about 17 dpa (keyway roots) and about 25 dpa (bores). Irradiation testing should be scheduled to keep ahead of the reactors as soon as possible.

- As long as it remains unconfirmed that creep is independent of the graphite grade, creep studies should be performed on AGR graphite.
- The fast flux densities in the material test reactors under consideration are about 12 dpa per year (HIFR ORNL), 6 to 7 dpa per year (HFR Petten), and 4 to 5 dpa per year (Idaho NL). Time for intermediate characterisation of irradiated samples and assembling/dissembling the irradiation rigs will be needed in addition.
- According to the current MTR experiments 8% weight loss per year are achievable.

AVAILABLE MATERIAL TEST REACTORS

- High Flux Reactor, HFR, at *Petten* (NL): Light water moderated and cooled 45 MWth tank-in-pool multi-purpose testing reactor, operated for 10 reactor cycles or 270 days per year, achieving a fast fluence of 0.835 dpa per cycle.
- High Flux Isotopes Reactor, HFIR, at ORNL (USA): Pressurized, light-water-cooled and -moderated reactor, peak fast flux density $11 \cdot 10^{18}/(m^2 \cdot s)$ (E>0,1 MeV), max. fluence 11 dpa/a.
- Advanced Test Reactor, ATR, at INL (USA): PWR type reactor, cooled and moderated by light water, with a Beryllium reflector, peak fast flux density $5 \cdot 10^{18}/(m^2 \cdot s)$ (E>1 MeV), 250 MW total power.

IMPORTANT ISSUES FOR FUTURE CREEP EXPERIMENTS

With respect to the number and complexity of factors influencing irradiation induced creep a "next" creep experiment cannot be designed straightforwardly. On the other hand the boundary conditions outlined in the previous chapter as well as feasibility aspects must be accounted for. Therefore, a list of aspects that need consideration may be helpful:

- As far as different stress modes are concerned experiments in tension as well as in compression are needed to see the difference. Nevertheless, there is a higher risk of sample failure in tensile creep experiments. In any case the level of stress should not be too high.
- Uniaxial as well as biaxial stress needs consideration. Significant difference, if any, should be looked for beyond turn-around. The need to have biaxial testing may depend upon the success of modelling creep based on CTE.
- Sample geometry should not use sharp notches to avoid stress concentration.
- Simultaneous oxidation makes creep experiments even more complex. Alternative ways to investigate the influence of weight loss due to radiolytic oxidation could be using samples pre-oxidised out-of-pile, installed samples or samples trepanned from real AGR bricks. However, pre-oxidised samples have not accumulated their weight loss under stress, and the question about differences between pre-oxidised and *in situ* oxidised material cannot be answered. The same is valid for installed samples (unknown history?), whereas trepanned samples normally have seen realistic conditions with respect to irradiation damage (available are samples with 14 to 16 dpa), stress and weight loss, but their history may not be well-known, important pre-irradiation data may not be available and intermediate property data do not exist.
- Even if pure technical difficulties and not scientific challenges are the driving force for the current activities on irradiation-induced creep of graphite, developing

new and better creep models is actually a scientific problem. One important requirement is to have "full" characterisation of creep and reference samples yielding as much as possible information about the unirradiated material. Next, subsequent irradiation steps and the respective loading of the irradiation capsules must be optimised in order to yield a maximum of information (data points) about the property changes at intermediate fluences. After the final irradiation to the target fluence again a "full" characterisation including destructive measurements such as strength tests and annealing has to be performed. In case that creep models to be developed make use of the irradiation behaviour of pure crystalline graphite highly-oriented pyrolytic graphite (HOPG) must be tested as well.

LAYOUT OF A FUTURE CREEP EXPERIMENT

Finally the discussions during the recent Workshop were directed to designing a next creep experiment. It was concluded that:

- A <u>main experiment</u> shall:
  - o be under compressive load;
  - o be in oxidising environment as well as in inert environment;
  - o contain virgin material (first choice AGR graphite followed by NBG-18);
  - o be taken to a fast fluence of 25 dpa;
  - o use a stress level between 5 and 10 MPa;
  - o be performed in the Petten HFR at a temperature close to 500°C.
- Several <u>satellite experiments</u> shall back up the main experiment and cover aspects that cannot be covered by the main one, mainly to:
  - o use tensile set-ups along with compressive ones;
  - o address creep reversal and creep recovery;
  - o use pre-oxidised samples;
  - o realise (in addition to 500°C) higher operating temperatures to validate models over a larger temperature range, and to achieve collaboration with other projects.

Details such as the sample geometry, distribution of the samples over the available irradiation volume (loading plan), determination of fluences, controlling of irradiation temperatures, sample preparation, pre- and post-irradiation characterisation programs *etc.* need to be discussed with test reactor laboratories.

References
Brocklehurst, J. E., Harrison, J. W., Kelly, B. T. and Martin, D. G. (1976). Report prepared under contract for KFA Jülich.
Brocklehurst, J. E. and Kelly, B. T. (1989). Report ND-R-1406 (S), Springfields Nuclear Power Development Laboratories.
Davies, M. A. and Bradford, M. R. (2008). Irradiation-induced creep - theories and models, Creep Workshop, Broadway, UK, July 21-23, 2008
Haag, G. (2005). Properties of ATR-2E Graphite and Property Changes due to Fast Neutron Irradiation, Jül-Report No. 4183, ISSN 0944-2952
Kelly, B. T. and Brocklehurst, J. E. (1977). UKAEA Reactor Group studies of irradiation-induced creep in graphite, *J. Nucl. Mat.* 65, 79-85.
Kelly, B. T. and Burchell, T. B. (1994). The analysis of irradiation creep experiments on nuclear reactor graphite. *Carbon*, 32, 119
Kennedy, C. R., Cundy, M. R. and Kleist, G. (1988). The irradiation creep characteristics of graphite to high fluences. International Conf. on Carbon and Graphite, Newcastle, UK.
Simmons, J. W. H. (1965). Radiation damage in Graphite (Pergamon, Oxford).

Part E

Plant Performance

Continuing Development of the Quarter Scale Core Model to Support the Safe Performance of AGR Cores

Anthony R. Roscow, Jim Skelton and Neil McLachlan

AMEC Nuclear Limited, 601 Faraday Street, Birchwood Park, Warrington, WA3 6GN, U.K
Email: anthony.roscow@amec.com

Abstract
The operation of British Energy's fleet of Advanced Gas-cooled Reactors (AGRs) depends on securing safety cases for their graphite cores. This requires ongoing research to support assessments of the functionality of the AGRs through life. The ageing effects in an AGR graphite core manifest themselves in a variety of ways including the distortion, and ultimately, the cracking of fuel bricks. AMEC Nuclear has built a quarter scale replica of an AGR core that allows the effect of brick cracking on the overall geometry of the core to be explored. The problem of measuring the displacement of individual fuel bricks and the distorted channel shapes has been addressed by using an automated machine vision system in which a camera scans over the channels and captures images of targets fitted to all the bricks. The images are processed using pattern recognition algorithms and the target positions resolved. To prove the system, a series of tests have been carried out using a 10x10x5 rectangular array.

Keywords
Machine-vision, Testing, Graphite

INTRODUCTION

British Energy is developing the safety cases for the AGR graphite cores to demonstrate that they can be operated with cracked fuel bricks. This is being done in stages to allow for the different types of cracked brick scenarios which might arise, in terms of the type of crack and their distribution throughout the core. Cracked bricks permit, and may even cause, greater relative displacements of the bricks or brick halves, and excessive displacements of these components could compromise fuel and control rod movements.

Tests using physical models of the AGR cores have always been integral to the validation of these safety cases. While it is accepted that these models may not behave identically to the graphite core, they provide an important means of validating the finite element (FE) codes and, where sufficiently representative, can be used to obtain information of direct applicability to the cores.

DEVELOPMENT OF THE QUARTER SCALE RIG

Based on experience gained working with the 1/8th scale whole core rig and full scale part core test arrays such as the IMC 3D rig, a quarter scale whole core rig has been built (Castro, 2006). Quarter scale was chosen because it provided the best compromise between scaling the key/keyway clearances and ease of manufacture and operation (actually 0.27 scale was used due to other design considerations).

Being the lead stations, the HPB/HNB design was considered as the prime candidate for this model. However, due to the number of small core components in particular in the presence of an inter-layer keying system, this geometry would have been very difficult to build, hence the simpler HRA/HY1 design was chosen (which could be subsequently modified to be similar to that at HPB/HNB). Even so, the core geometry had to be simplified further such that the whole core could be built using the minimum of specific components, for instance, only one design of "fuel/ reflector" brick and interstitial brick were used. These bricks were based on the geometry of components in the central layers of the cores but without rocking features (pivots on the brick base). Furthermore the method of restraining the edge bricks was simplified to make the outer columns completely fixed. These features were not relevant to the initial test programmes. The four quarter scale components are shown in Figure 1 ("Fuel/ reflector" brick, cracked half fuel brick, interstitial brick and loose key).

The quarter scale core rig is purely a geometric model, used for static testing and modelling the quasi-static end conditions. It does not model the transient thermal expansion or contraction of the core, nor does it model dynamic effects, hence no attempt was made to represent density or friction. Therefore the most important material property was dimensional stability, leading to the choice of extruded and machined aluminium for the bricks and stainless steel for the loose keys.

FIGURE 1: Quarter scale core components.

The key/key-way clearances are important parameters that affect the movement of the bricks within the core. Therefore to minimise pessimisms in the test results, particular attention was given to their reproduction, such that even their tolerances were scaled. This required that the bricks were manufactured using state of the art techniques.

FIGURE 2: The quarter scale rig in operation.

As with most rigs of this type, the important parameter in the design of the restraint frame was stiffness. The frame was required to house 5712 fuel bricks and 5148 interstitials, in an array 24 fuel bricks in diameter and 12 layers high, weighing a total of 35 tonnes, yet deflection of the structure had to be less than 0.5mm. This resulted in the frame being built from 533mm deep beams which, when loaded, are only subject to 2% of their ultimate strength. The octagonal frame is tilted by a pair of electric jacks in any one of eight directions, allowing lateral loads to be imparted in all cardinal, *i.e.* North, South, *etc.* and oblique *i.e.* North West, South East, *etc.*, directions. The rig is shown in operation in Figure 2.

Measurement System

Obtaining accurate measurements from potentially over 5000 core bricks presented a significant technical challenge. Fitting discrete instrumentation, to even a small proportion of them, would have been both expensive and impractical. However building on British Energy's investment in on-load core monitoring systems, AMEC Nuclear had already developed a machine vision system for measuring the profiles of fuel channels, which could be adapted for this new task.

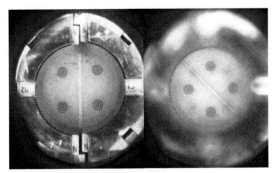

FIGURE 3: Typical images from the camera system.

Targets printed on transparent film were fitted inside the bores of the intact and doubly cracked fuel bricks. Under computer control, a camera scanned over the top of the array, stopping over each fuel channel. By deliberately using a lens with a shallow depth of field, individual targets on each layer could be focussed in isolation and the images captured. A machine vision system then processed the images and determined the positions of the targets.

Camera motion control systems and pattern recognition software capable of the task can be bought off the shelf, but the key to success, as in any machine vision project, is to develop a consistent diffuse lighting system. Incident lighting was rejected because specular reflections would cause artefacts in the image, that could not be removed by image processing. Instead, a back lighting system was developed. The replica moderator support structure (the base that locates the base of the bricks) was made from a sheet of specially engraved acrylic to act as a diffuser and the light was injected into the edges of the sheet using a ring of 96, 5W light emitting diodes. Again the key to obtaining images with good contrast is eliminating extraneous light, so the camera was fitted with a shroud that extended to the surface of the array. Figure 3 shows typical images from this system.

QUARTER SCALE RIG OPERATION

The procedure for carrying out a test with the rig is to set up a specific configuration of cracked bricks within the core. The whole core is tilted to 30 degrees in any one of eight directions, four cardinal (N, E, S, W) and four oblique (NE, SE, SW, NW). This imparts a static lateral force on all the bricks and forces the channels to their extreme positions. To eliminate uncertainties in the initial positions of the bricks, such as would be introduced during the build or left over from the previous tilts, many of the tests were carried out in opposing pairs (*i.e.* a South tilt immediately followed by a North tilt) and the difference in displacement considered.

Mini Core Tests

Operating the quarter scale rig can be expensive. It is estimated that completely reconfiguring the clusters of doubly cracked bricks within the whole 24 brick diameter, 12 layer array could take weeks, so it is essential that the development of the test programme and the changes between configurations are well planned. Hence, to commission the whole measurement system and analysis procedure, and to obtain interim results for validation of computer models, a simpler configuration was chosen.

FIGURE 4: Image from the top of the mini-core.

A stiff steel box was built and braced within the restraint frame. This was filled with 500 fuel bricks in different configurations, arranged in a 10x10x5 array. This mini-core was restrained by the brick end face keys at the base and keyed in the walls of the box by nylon spigots, replicating the action of the integral keys. This array was not intended to represent a full reactor core, and the behaviour of the components could only be indicative of what might happen in an actual core, but the small number of components and the straightforward geometry meant that the measurements could be easily correlated with the output from simplified FE models. The layout of the mini-core is shown in Figure 4.

Test Configurations
Four test conditions were chosen spanning the full range of doubly cracked bricks:

- Fully intact bricks.
- 100% doubly cracked, with random crack orientations in plan but aligned within in each channel.
- 50% doubly cracked, with random crack orientations in plan but aligned within in each channel.
- 50% doubly cracked, with random crack orientations throughout the core.

These were tilted in a series of oblique and cardinal directions, a total of 24 separate tests. In addition some tests were carried out to measure the performance of the measurement systems and the deflection of the frame.

Interpretation of the Test Data
While gross displacements of the bricks are significant parameters because fuel stringers or control rods could become trapped in grossly distorted channels, other important parameters include shear and separation of the doubly cracked brick halves.

The large amount of data collected, potentially up to 2000 coordinate pairs, and the different parameters that can be calculated from these, means that statistics are the best approach to interpreting the test results. Frequency distributions, specifically cumulative distributions - because the shape of the curve is not dependant on class interval, allows the test data to be readily compared with the output from the FE models.

Maxima and minima of the measured parameters are important because they indicate the possibility of certain safety and functionality criteria being exceeded such as distorted fuel assemblies, trapped fuel stringers, or failed control rod insertion. However these values are strongly affected by rogue data, such as damaged targets or misidentification by the pattern matching software. For this reason, the 2nd and 98th percentiles of the distributions were calculated and were regarded as more representative of the real maxima and minima.

In the cardinal directions, forces are transmitted between bricks through the integral keys on the interstitial bricks, and in the oblique direction, the forces are transmitted through the loose keys. As stated above, the movement of the core is strongly influenced by the scaled key/keyway clearances which are 0.18 mm for the integral keys and 0.06 mm for the loose keys. Hence the difference in clearances suggests that the core has the potential to behave anisotropically.

Gross Displacement of the Bricks in the Test Rig
Despite the anisotropy of the core, the gross displacements of the bricks, and brick halves, were similar for both the cardinal and oblique tilt tests. In both cases the magnitude increased with the percentage of doubly cracked bricks but the relationship with layer number was less well defined. The largest displacement was in layer 3. In the oblique direction the largest movements were in layers 3 to 5. In all tests movement was found to be predominantly in the tilt direction, as would be expected.

Test Rig Cracked Brick Shears and Separations
Again the shear separation results from the cardinal and oblique tests were similar, with the shear increasing with the percentage of doubly cracked bricks. At 100% doubly cracked, the

largest shears were found to be in layer 5 for a cardinal direction tilt and in layer 3 for an oblique tilt. Generally the shears for layers 3, 4, 5 were of similar magnitude. The largest separations were in layer 4 for cardinal and oblique.

The cracked brick components are idealised, in that the cracks are modelled as smooth plane surfaces, exactly 180° apart. In real cores, the cracked surfaces would be highly irregular, and whilst this would probably not affect separation, it might significantly impede shear.

CONCLUSIONS

Under the sponsorship of British Energy, AMEC Nuclear has developed a physical model of a full core, more representative than any previous rig. The design has successfully integrated technologies from previous core projects and provides a useful tool kit for validation of the safety cases.

The development of the rig and the rig components had proved the versatility of the design and the validity of the approach. The machine vision system has been shown to be a reliable method of collecting data, and where anomalies occur due to damaged or misidentified targets, a sensible approach to data processing can help to compensate.

The edge restraints for the mini core are not yet representative, and this latter point is being addressed in the next phase of the project, as the rig is being prepared to accept a full core, more representative restraints have been designed and installed, the conceptual design and the installation are shown in Figures 5 and 6.

FIGURE 5: Conceptual design of the quarter scale restraint.

FIGURE 6: Installation of the quarter scale restraint showing the acrylic base.

The tests on the mini-core do not demonstrate any significant anisotropy, but this core is less than 10% of full core, and differences in the behaviour in cardinal and oblique tilts may be apparent when the full core testing is under way. The shears and separations of the doubly cracked bricks were found to increase with layer number and with the percentage of doubly cracked bricks, with the maximum values being obtained in the 100% cracked tests. The gross motion of the bricks components also increased with the percentage of doubly cracked bricks but the relationship with layer was less well defined.

This project has demonstrated that, even to deliver what is only a small part of the whole Safety Case, a large team of people with a very wide range of skills is required, from craftsmen manufacturing and assembling the rig, the core and test engineers designing the experiments, through to the analysts perfecting the FE models.

Acknowledgements
The authors would like to thank British Energy Generation Ltd for permission to present this paper. The views expressed in this paper are those of the authors and do not necessarily represent those of the British Energy Generation Ltd. Special thanks to Dr S. J. Pountney in assisting with the experimental programme and its reporting.

References
Castro, E. (2006). Investigation of degradation of AGR graphite core geometry using a whole core scale model. *Proceedings of the international youth nuclear congress 2006, Stockholm/Olkiluoto 18-23 June 2006 332.1-332.8*

The Effect of Reactor Parameters on AGR Refuelling at Hinkley Point B

Christopher J. Wallace[a], Graeme M. West[a], Stephen D. J. McArthur[a], Dave Towle[b] and James Reed[b]

[a]Institute for Energy and Environment, University of Strathclyde, Glasgow, UK, G1 1XW
[b]British Energy, Graphite Core Project Team, Barnett Way, Barnwood, Gloucester, GL4 3RS
Email: cwallace@eee.strath.ac.uk

Abstract
The automated analysis of refuelling data to enhance condition monitoring has recently been introduced at some of British Energy's Advanced Gas-cooled Reactor (AGR) stations. It has been proposed that various reactor parameters, such as temperature and pressure, could affect the measurement of the load of the fuel stringer during insertion or removal from the reactor core, which is recorded as part of the Fuel Grab Load Trace (FGLT). The influence of reactor parameters is not currently considered as part of the automated analysis due to limited knowledge of the strength of any such effects. This paper discusses work undertaken to fuse available FGLT data from the Hinkley Point B station with reactor parameter data from station logs to detect whether any correlation exists. After extracting station logs and matching the appropriate parameter values to a specific refuelling event it was found that there was a correlation between the dome differential pressure and deviation in the measured load of the fuel inside the core region relative to the absolute fuel stringer weight, of up to 100kg. It was also found that there was no detectable correlation between deviations in the measured load and the vessel pressure, temperatures of various core regions or inlet guide vane positions. The detected correlation could potentially conceal or exaggerate the identification of features in the automated FGLT analysis, which uses the anomalies in the FGLT to detect fuel channel distortions. In the context of securing the safe operation of AGRs, these results will be factored into future automated analyses of FGLT to produce more accurate descriptions of channel distortions. Furthermore, increased understanding of the current condition of the core will help strengthen the safety case.

Keywords
Condition monitoring, Data fusion, AGR

INTRODUCTION

Recent work to enhance condition monitoring at British Energy's (BE) Advanced Gas-cooled Reactor (AGR) stations has included analysis of fuel movement data to detect distortions in the fuel channels. During each refuelling event, when a fuel stringer is inserted or removed from the core, a measurement of the load and height is recorded to ensure that the fuel moves freely during charge and discharge and is correctly seated in the core. The fuel stringers have stabilizing brushes which guide the assembly through the core. By considering that the measured load of the fuel as it passes through the channel is affected by frictional components within the core, it should be possible to detect channel distortions from analysis of the FGLT.

The British Energy Trace Analysis (BETA) system, developed at the University of Strathclyde, West *et al.* (2006), performs an automated analysis of the FGLT using envelopes of expected behaviour generated by the system and incorporates core damage profiles produced on a test rig, Skelton (2005). BETA is currently installed within BE to complement the manual analysis of FGLT.

With the ability of some AGR stations to refuel on partial load, certain core conditions must be met before refuelling can commence. It has been proposed that variation in these reactor parameters, even within the strict refuelling criteria, could affect the measured load of the fuel stringer on the FGLT. If this was the case, there is concern that such effects, if not properly compensated for, could potentially conceal or exaggerate features within the FGLT, limiting the accuracy of the automated analyses of BETA. This paper describes the analysis of these reactor parameters in relation to the FGLT and attempts to determine if any correlation exists.

As part of a more general theme, the work conducted here is an example of the use of data fusion to combine disparate sources of data to form a better description of the core region. It will be seen that the main issue in performing such fusion is not the analysis of the data, but the acquisition, storage and verification of the data. This problem is a result of the age of the current AGR stations, and the technical limitations of the existing IT infrastructure.

Reactor Data
Each AGR station maintains logs that contain a selection of reactor parameters including temperatures and pressures from various points within the core region. These include: *Dome Differential Pressure* (the difference between the gas circulator inlet pressure and the circulator outlet pressure), *Vessel Pressure, Graphite Temperature, Gas Inlet Temperature, Gas Outlet Temperature* and *Inlet Gas Vane Position* (which controls the gas flow rate into the core). These parameters are gathered and utilised in existing core condition monitoring applications. They are also recorded as part of station records and archived on optical discs.

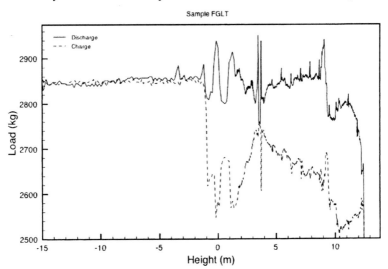

FIGURE 1: An example FGLT from Hinkley Point B. The peaks and troughs in the load are indicative of the internal structure of the channel and the varying frictional forces caused by changes in channel diameter relative to the brushes that separate the fuel from the channel surface.

Fuel Grab Load Trace

The FGLT has traditionally been recorded on paper traces to ensure that the fuel safely traverses the core and is properly seated, but recently electronic loggers have been installed in some stations alongside the paper trace recorders. A large number of historical FGLT have since been electronically scanned in order to construct a model of known behaviour of AGR cores. Figure 1 shows an example of a FGLT at Hinkley Point B. As the FGLT measures the load of the fuel stringer during insertion and removal, the FGLT effectively measures the frictional components acting on the stringer as it moves through the channel. The peaks and troughs in the FGLT represent structural features inside the channel, such as the interface between brick layers or transitions between different sections of the core. These alter the frictional forces acting on the brushes on the stringers which in turn affect the measured load. Note that the 'height' described here is actually the depth (from 0m) into the core region, and the negative height values indicate that the stringer is in positions above the core, inside the charge machine. When the fuel is seated in the bottom of the core, the load quickly falls to zero as the fuel is fully supported by the core. Analysis of the FGLT provides a way of detecting channel distortions if characteristic features can be identified which are indicative of cracking of the graphite bricks.

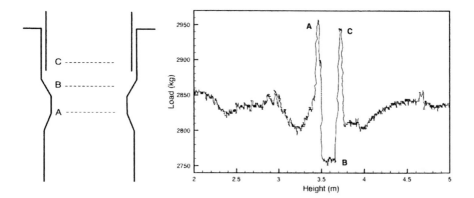

FIGURE 2: A simplified schematic of the channel constriction used as the point of reference for this study on the left and on the right the section of the FGLT that describes this passage. The features visible in the FGLT are due to the nose brush at the top of the fuel stringer passing the constrictions in layer 12 of the channel. Note in particular the constrictions at A and C which correspond to the characteristic 'double spike' feature of the discharge FGLT and the wider section of the channel at B represented on the FGLT as the minimum.

As most AGR stations approach the end of their predicted operational lives, the FGLT is also being used to compare the historical state of the reactor core with the current state, in order to develop a knowledge base of what normal behaviour is. The ability of BE to assess the state of the graphite core is important the safety case, as the state of the graphite core determines the safety and viability of continued refuelling. A key aspect of this work, which relates to BETA is the generation of envelopes of expected behaviour, which can be used to more accurately detect anomalies in the FGLT.

METHOD

In order to determine whether there is a consistent effect on the load caused by one or more reactor parameters, a point of reference is required. The point selected is the load minimum, shown as point B in Figure 2. This value of load corresponds to the nose brush on the fuel stringer after passing a constriction into a wider section of the channel where the fuel stringer is suspended in the channel without the brushes coming into contact with the channel wall. This data point is selected as a load measurement composed only of the absolute weight of the stringer plus the effect of any reactor parameters. Comparison of this load measurement with the absolute load of the fuel stringer allows the effect of any reactor parameters to be isolated.

Point B was identified by height of the load measurement in the channel, specifically the average load value between 3.55m and 3.65m. Point A was identified as the load maximum at heights greater than 3m, but prior to point B. Similarly, point C was identified as the largest load value after point B, but at less than 4m. These values were calculated using spreadsheet software to import the FGLT data, then manually ensuring the appropriate data range had been used.

The absolute load of the fuel stringer, $Load_{abs}$, was taken to be the weight of the fuel inside the charge machine, specifically the average load between heights of -10m and -15m in Figure 1. Subtracting of the stringer at point B, $Load_{Core}$, from the load of the stringer inside the charge machine, $Load_{abs}$, a measurement of the difference of the measured load of the fuel inside and outside of the reactor can be made.

This simple calculation is performed for 240 FGLT from HPB between 2005 and 2008. The calculated value was found to be between 10kg and 120kg, that is $Load_{Core}$ is less than $Load_{abs}$, implying that the effect of reactor parameters is to apply an upwards force, acting against the weight of the stringer.

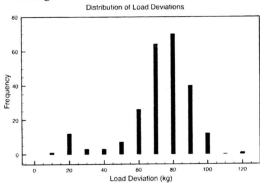

FIGURE 3: A statistical distribution of load deviations calculated from Hinkley Point B refuelling events.

Figure 3 shows the distribution of the calculated values of $Load_{Core}$. The small peak around 20kg is composed primarily of data from an outage, where the calculated load deviation was much smaller than those at normal on load operating conditions, which are contained within the larger peak around 80kg. This distribution shows that most refuelling occurs at similar operating conditions, however, there is enough variability in the load deviations that a common value of load deviation cannot be confidently used.

Data Processing

In order to determine whether a correlation exists between any reactor parameters, a single average value for each parameter over the course of an event is required. Recent work within BE has shown that the parameters vary by less than 7% on average during a refuelling event. Assuming a similar value for other parameters, a single average value can be used with good confidence for each event.

The station logs were extracted from the optical discs for time periods during which refuelling events were known to have taken place. The log data was then compiled into a database, the structure of which was based on the parameters and timestamps. The reactor parameter data from HPB is sampled at a rate of one measurement per minute, which resulted in a database size of around 120,000 rows.

The FGLT logger and the reactor parameter data recorder systems are separate within the station and as such are not synchronised. This presents a problem in terms of relating the appropriate reactor parameter data to the correct event, specifically the times associated with the data. This problem is simplified somewhat by the procedural requirement that refuelling occurs at DDP less than 0.65 bar, which, after applying an appropriate filter to the database reduces the number of rows of data to around 60,000.

Each AGR station maintains a log of refuelling events that contains a list of which channels were refuelled, on what date, and the time that refuelling began. This time was used to identify the approximate start time of the refuelling event within the reactor parameter data. Using this time as a guide, the data was further subdivided into individual refuelling events, and regions of reactor parameter data were matched to events in the refuelling log.

Prior to refuelling taking place, the reactor power is reduced to less than $450MW_T$, the vessel pressure to between 37bar and 39.5bar and the DDP to below 0.65bar[11]. After discussion with BE staff it was considered that the first time that the DDP falls below 0.65bar, it can be assumed, with reasonable confidence, that refuelling has begun.

Given that a refuelling action normally lasts one hour (approximately 30 minutes for charge, or insertion of fuel and 30 minutes for discharge, or removal of fuel), the average reactor parameter value over the first 30 minutes was assigned to the discharge and the average reactor parameter value over the last 30 minutes before the DDP rises above 0.65bar was assigned to the charge. The period over which the DDP remains below 0.65bar is slightly longer than an hour, but this is due to period between disposing of the spent fuel and lifting the new fuel. This process was repeated for each fuel movement for which both reactor parameter data and FGLT were available.

RESULTS

Once each refuelling event had been assigned a set of reactor parameter values, the load deviation for each event was as a function of each reactor parameter in order to detect whether any correlation exists. The measured load deviation was compared to each of the

---

[11] Note that these values are specific to Hinkley Point B (HPB) and can vary between stations.

reactor parameters: *Dome Differential Pressure, Vessel Pressure, Graphite Temperature, Gas Inlet Temperature, Gas Outlet Temperature* and *Inlet Gas Vane Position.*

Only for the DDP was a positive correlation detected between measured load deviation and parameter value. The other reactor parameters exhibited no obvious correlation, an example of which can be found in Figure 4. The vessel pressure, the pressure within the core, is plotted against load deviation in Figure 4 for both charge and discharge[12]. The largest grouping of data points reflect the core at normal operating conditions, whereas the scattered data points at lower pressures occurred during an outage.

The other reactor parameters, when compared to the measured load deviations, exhibited similar distributions; namely a large grouping of values that reflect normal operating conditions and a small number of points recorded during outage out with the main group, neither of which showed any identifiable trend.

Figures 5 and 6 show the correlation between the DDP and the measured load deviation. In both cases, the distributions suggest a linear relationship. A linear regression of each data set generated a function for each and correlation coefficients of 0.479 and 0.888, respectively. The correlation coefficients are a measure of how closely the function that best approximates the relationship between the variables matches the real data values. In this context it, the function that describes the discharge data has a better fit than that of the charge.

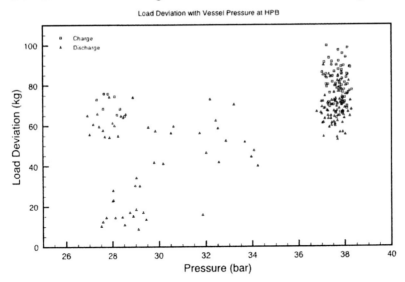

FIGURE 4: The variation of vessel pressure with calculated load deviation. There is little variation in load deviation relative to vessel pressure and no discernable relationship.

---

[12] The charge and discharge data are presented together, for efficiency, in Figure 4 where no relationship was detected, but are separated in the case of a positive relationship - Figures 5 and 6. This was done because the stringers are often damaged or distorted after extended periods in the core, and it is possible that any effect from a reactor parameter might affect stringer types differently.

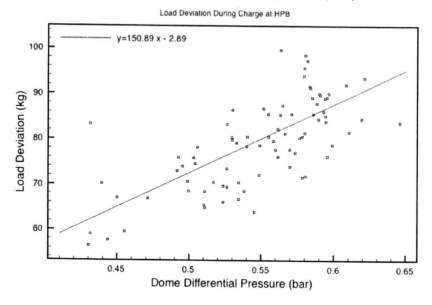

FIGURE 5: A linear regression is fitted to the calculated load deviations plotted against the DDP at which they were recorded. The calculated correlation coefficient was $R^2=0.479$.

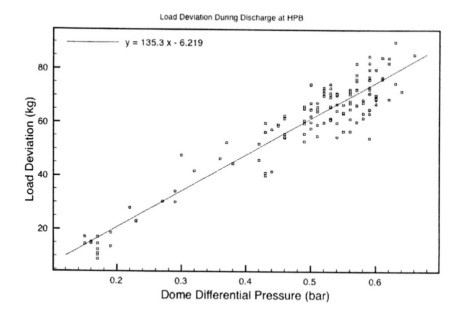

FIGURE 6: A linear regression is fitted to calculated load deviations plotted against the DDP at which they were recorded. The calculated correlation coefficient was $R^2=0.888$.

CONCLUSIONS

These results quite clearly indicate a relationship between DDP and load measurements during FGLT that requires consideration on its impact on the FGLT analyses performed by BETA. Further work, by means of a similar study at another station, is required to determine whether the effect is consistent across all stations or if it varies in strength.

The fact that a positive load deviation was found in every case, that is the fuel stringer appears to weigh less inside the core than outside, suggests that the effect of the DDP is to produce an upwards pressure that supports some of the weight of the fuel stringer. This is consistent with the discovery that larger DDP result in larger deviations. It is also possible that with better matching of FGLT event to reactor parameter time, the correlation between DDP and load deviation might have been more distinct.

The effect was found to be more consistent in the discharge data, however, this was due to a lack of data at low pressures (*i.e.* from outages) in the charge data. During refuelling on outage there can often be several weeks between the removal of fuel and the insertion of new fuel, which was not accounted for when the reactor parameter data was extracted from the optical discs. Outlying data points are likely to be the result of the average reactor parameter value calculated for each event deviating slightly from the actual reactor parameter value where the load minimum occurred inside the core.

When the consistency of the effect is determined, it will be straightforward to implement an appropriate function in BETA to remove the effect before analyses are performed. This, however, relies on the availability of DDP data, which is not currently attached to refuelling events. Work is underway between BE and the University of Strathclyde to implement a data fusion scheme to combine multiple sources of reactor data, which should facilitate the use of DDP data within BETA.

While there are strong arguments to be made for higher quality analyses that can be conducted with data from multiple sources, a significant amount of preparatory work is generally required to achieve this. This problem is primarily due to the age of AGR stations, and the lack of a properly integrated IT system. For the purposes of this work, the filtering of the appropriate data was performed using commercially available software and assembled by hand, but for regular or larger scale analyses of this sort, a customised software solution is required. The University of Strathclyde is working closely with BE to develop the use of advanced computing methods, such as intelligent agents and machine learning, for data management and condition monitoring. These techniques will help allow more detail, consistency, and efficiency in existing analyses as well as a greater ability to detect new relationships by analysing more data than was practical before.

Acknowledgements
The authors would like to thank British Energy for financial support of this work. Thanks go in particular to Graham Buckley and Ray Peacey at British Energy for their help and expertise. Please note that the views expressed in this paper are those of the authors and do not necessarily represent those of the British Energy.

References
Skelton, J. (2005). Fuel grab load trace applied to the detection of cracks in AGR fuel bricks: Final report on the experimental and analysis programme.
West, G. M., Jahn, G. J., McArthur, S. D. J., McDonald, J. R. and Reed, J. (2006) Data mining reactor fuel grab load trace data to support nuclear core condition monitoring. *IEEE Transactions on Nuclear Science*, 53, 3.

Part F

Graphite Reactor Decommissioning

Twenty-First Century Solutions to the Graphite Dismantling and Radwaste Problems

Anthony J. Wickham[13] and David Bradbury*

Nuclear Technology Consultancy, PO Box 50, Builth Wells, Powys LD2 3XA, UK
* Bradtec Decon Technologies, The Wheelhouse, Bonds Mill Industrial Estate, Stonehouse, Glos GL10 3RF, UK
Email: tony@tonywickham.co.uk

Abstract

There is an increasing level of concord internationally that the process of decommissioning old reactors should be speeded up, on socio-political grounds and because of fears of the loss of knowledge and expertise over longer timescales. On a purely financial basis, there remains however an argument for doing nothing, and thus the removal of graphite blocks from a previously intact pressure vessel may in this latter case be regarded as a retrograde step. The planning and execution of suitable final disposal routes for graphite waste has not been satisfactorily addressed and, where blocks have been removed or such retrieval of intact blocks is planned, removal to a temporary surface facility may be the only present option. Graphite is a unique waste form, being essentially elemental carbon, with a well-understood chemical form and chemical/physical/mechanical properties which certainly justify its consideration as a separate waste stream with the potential for individual treatment. Such treatment can begin with its removal from the vessel, where mechanical dismantling of a core as intact blocks may not necessarily be the easiest or cheapest option. There then follows a potential for pre-treatments which enable concentration and containment of isotopes which have potential for adverse effects in the environment. Even isotope dilution should not be dismissed as a possibility because such dilution can enhance radiological safety if coupled with containment. There are numerous alternatives to repository burial which, on the basis of an objective risk assessment, appear to have considerable technical merit in addition to potentially large cost savings. This paper proposes that it is time to consider "waste hierarchy" solutions for graphite management rather than following a one-dimensional approach of intact block retrieval, encapsulation, storage and final burial.

Keywords
Irradiated graphite, Radioactive waste, Decommissioning

INTRODUCTION

The debate about the appropriate strategy to decommission graphite-moderated reactors and to deal with the waste graphite has been continuing for at least thirty years without resolution. The original UK proposals for long-term storage of the graphite within the pressure vessels (up to 135 years), which were mirrored by French and Italians plans, albeit for differing periods, and which had tended to be assumed by the countries of the Soviet Union at that time for the RBMK reactors (essentially in these cases there is no pressure containment for the graphite, and no plan either), have been slowly overturned by social, political and regulatory pressures into planning for more rapid dismantling. Unfortunately, the plans for graphite disposal, which have essentially considered only a deep repository solution until recently, have been hampered by the lack of any country to create such a repository, leaving temporary surface storage as a necessary, but inappropriate alternative.

---

[13] Author to whom any correspondence should be addressed.

A position has been reached where we now have some reactors completely dismantled (GLEEP in the UK and Fort St. Vrain in the USA, with the GLEEP graphite destined for calcining to reduce the content of ^{3}H and ^{14}C whilst that of the latter, in the form of fuel blocks, remains in a storage facility (Fisher, 1998); some from which the graphite is fully and successfully removed (WAGR in the UK) to temporary storage; and plans for a more accelerated dismantling schedule (UK Magnox plant, French UNGG and Lithuania's RBMKs). In the case of UNGG plant, it is intended that the later reactors will be dismantled under water, but earlier designs cannot be treated this way. There are no plans, nor any necessity seen, for dismantling underwater for Magnox or AGRs. ANDRA, in France, has taken a lead in designing a relatively shallow repository specifically for graphite waste (Ozanam, 2008). A depth approaching 200 metres appears to be the currently favoured design.

Debate about the options available for treating the graphite has also taken place widely: White *et al.* (1984) for the EU, using UK Magnox reactors as an example; Marsden and Wickham (1998) compared options then available to UK and Russian operators; and perhaps the initial challenge to conventional thinking, made in the context of an objective risk analysis, was made by Neighbour *et al.* (2000). The ability of calcining, already mentioned, to eliminate high proportions of certain isotopes has been studied at the laboratory scale in detail using graphite from the German AVR, to obtain preliminary kinetic information (Fachinger, 2008).

Deep burial of a cumulative total exceeding 200,000 tonnes of irradiated graphite introduces issues of cost and long-term containment of isotopes which the last reference discusses alongside the presently outlawed option of sea disposal and the opportunity to simply incinerate the graphite. Powerful arguments about the minimal effect on global ^{14}C emissions of such an activity, in comparison with natural upper-atmosphere production rates from ^{14}N, are made by Nair (1983), and reinforce a growing suspicion that limits for isotope releases from disposal waste and associated treatment processes which are based on the linear no-threshold approach make little sense against the significant natural background of radioactivity which this planet has sustained for millennia. This, along with careful consideration about the impact of other sensitive isotopes arising from graphite, such as ^{36}Cl, encourages renewed consideration of other alternatives to the deep (or indeed the shallow) repository for this waste stream. These include pre-treatments to reduce isotope content before disposal, alternative process technologies (which include the retrieval options from the original stacks as well as destructive treatments of the graphite), and recycling options.

RETRIEVAL OPTIONS

Until very recently, all completed and planned dismantling of graphite stacks assumes mechanical lifting of some kind, and the transfer of intact blocks via some form of temporary storage to a final disposal site. Whilst alternative destinies to repository burial are clearly possible in this case, there is no reason to confine innovative thinking to the disposal stage.

Discussions between various bodies in the UK and USA, have resulted in the development of an entirely different dismantling strategy, best described as 'Nibble and Vacuum'. Here, scabbling techniques as used already to break up concrete are envisaged to break up the graphite blocks (and perhaps other components too) within the reactor vessels, before conveying the debris by suction tube to the next stage of processing. The potential financial savings of such an arrangement, which requires only simple engineering, are very large in comparison to conventional insertion of large lifting machinery. A very recent proposal (Rahmani, 2008) sees the crumbled material reduced in particle size and mixed with water to form a foam slurry which is then directly introduced to underground disposal in appropriate strata, spent oil and gas fields, and so forth. Deep boreholes for this option could even be drilled on site, potentially avoiding the need for transport. However, the present authors have been more concerned with chemical treatments, explicitly gasification. Originally, the concept was to perform the chemical reaction of graphite to form gaseous products within the reactor vessel, but the initial application of the 'nibble and vacuum' concept permits a much greater efficiency in the process, which is outlined below.

STEAM REFORMING / ISOTOPE SEPARATION

The concept, which can only be briefly described here, is based on the classical technique of 'steam reforming'. The graphite is reacted with steam between 700-1000°C to form a mixture of carbon monoxide and hydrogen, after which two options are available. One is re-deposit the carbon and make it available for new carbon-based products for nuclear applications (including new graphites for future HTR build, highly-absorptive carbon filters, electrodes for existing processes used on other forms of radwaste, and so forth) – effectively this is 'recycling' the bulk material. The second is to move to some form of isotope separation to remove the majority of the ^{14}C (other important non-volatile isotopes will remain in the small quantities of ash for conventional disposal whilst other volatiles such as ^{3}H and ^{36}Cl can be removed from the off-gases): this recovered ^{14}C has the potential to displace the need for newly manufactured ^{14}C from the irradiation of aluminium nitride in what is a surprisingly large world-wide market for this isotope. The largely depleted fraction, essentially slightly contaminated ^{12}C may, after conversion back to CO_2, can be compressed and removed by tankers, used in other ways, discharged within accepted limits, or sequestered. Dilution with CO_2 sequestered from fossil-fired plant would absolutely remove any residual risk from long-term dispersal by reducing of the specific activity to background levels. It should be noted that this is "dilute and contain" not "dilute and disperse".

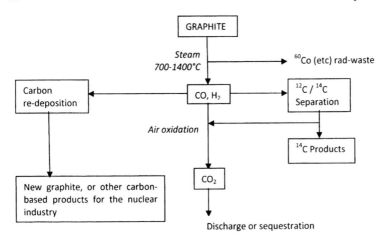

FIGURE 1: Chemical process diagram for steam reformation of irradiated graphite.

An important additional concept of the planning of the overall graphite removal and chemical processing relates to the other isotopes in the off gases. Although not shown in the diagram in Figure 1, a number of possibilities exist for their entrapment. [36]Cl is of particular concern to regulators, despite its exceedingly long half life (300,000 years) and the consequent low decay rate, because of fears of its long-term re-concentration in the environment and the food chain. Previous thinking has tended to concentrate on isotope concentration: however, it is now considered that serious consideration should be given to isotope *dilution* as means of lowering specific activity and thus assuring future radiological safety. This may be undertaken as well as, rather than instead of, burial and isolation. It is important that the diluent of non-active isotope should be in the same chemical form as the active component.

The outline of a complete dismantling and gasification process is shown in Figure 2, including the scrubbing of vapour-phase isotopes and the grouting of the secondary waste arising from ash: here, the potential for separation of ^{14}C from ^{12}C in the generated CO_2 is omitted for clarity, but this is discussed in the next Section.

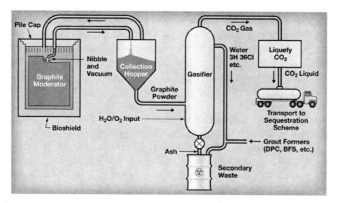

FIGURE 2: Simplified graphite-processing diagram.

CARBON ISOTOPE SEPARATION

We have already mentioned the ease of mobilisation of a large proportion of the ^{3}H and ^{14}C found on graphites from GLEEP and AVR simply by heating. The success of this is believed to depend upon the location of the isotope on the graphite. [3]H has long been known to diffuse to the pore surfaces of graphites and to have the ability to exchange easily with other hydrogenous impurities in reactor coolants (Best *et al.,* 1976). UK and German data suggests that a majority of the ^{14}C associated with stack graphite also resides on pore surfaces as carbonaceous deposits which have obtained their content of ^{14}C from a constantly-replenished source of impurity ^{14}N in the CO_2-based coolants, which is in accord with its obvious mobility upon calcination (Marsden *et al.,* 2002); this is supported by the very high ^{14}C content associated with RBMK graphite, which is exposed to a N_2/He cover gas. However, very recent calculations in support of French UNGG decommissioning suggest that all ^{14}C in graphite from Bugey 1 comes from either ^{13}C or ^{14}N within the graphite itself (Poncet, 2008), and this disparity is to investigated by an international collaboration in the early part of 2009.

Carbon isotope separation processes rely upon the exploitation of the small mass difference in various compounds of carbon to separate the inactive ^{12}C and ^{13}C (at approx. 1% natural concentration) from ^{14}C. In concentration terms, the amount of ^{14}C is tiny, and its separation therefore difficult, requiring many stages in conventional separation technologies such as the amine-carbamate reaction (see, for example, Oziashvili and Egiazarov, 1989, although very recent Romanian claims to have improved this process dramatically are under investigation). In addition to amine-carbamate, three other potentially promising technologies are currently under investigation:

- Pressure-Swing Absorption of carbon monoxide (the most efficient form in which to maximise the mass difference between the isotopes) using zeolite columns at sub-zero temperature: this technique was developed in Japan but requires further evaluation to establish its efficiency;
- Cryogenic distillation, also of CO (initially developed in Canada); *and*
- Gas centrifuge.

These alternatives have recently been reviewed by one of us (Bradbury, 2008) and are currently under further investigation as part of the CARBOWASTE project which is funded under the EU 7th Framework research programme with over 20 participating organisations in Europe and world-wide. In addition, an IAEA initiative is also under way to undertake targeted experimental work on some of these potential solutions.

We also note however that isotope separation is inappropriate for solutions which involve burial or other isolation, since this increases greatly the specific activity of the ^{14}C such that if escape from repositories becomes possible, the consequences can be greater. In these circumstances, dilution is better. However, the separation and concentration is necessary to permit recycling of graphite, or to collect ^{14}C as a precursor to labelled products required for scientific and medical purposes. It is therefore entirely possible that *small-scale* isotope separation will be required, rather than treating the graphite from entire stacks. In this case, a wide choice of technologies appears to be ready for development.

ALTERNATIVE OPTIONS

Pre-Treatments

A recent evaluation of available data on the leaching of radio-isotopes from graphite (Wickham, 2008) confirms that leaching of most isotopes is largely complete within a period of about 450 days, although the leaching of ^{14}C and ^{3}H is surprisingly slow, the former appearing in some ways to support the recent conclusions of (Poncet, 2008) that the release requires solid-state diffusion from within the carbon lattice. Acidic solutions, used in a limited number of tests, appear to accelerate significantly the release rates of most isotopes, and this offers a potential for a clean-up process of graphite which might have the result of lowering its waste categorisation for repository disposal. The use of surfactants and sequestering agents to facilitate such a process is to be investigated at The University of Manchester, UK. Such a process should certainly be considered alongside the calcination option for pre-treatment, and alongside other options for graphite disposal, in order to reach optimal solutions for particular situations.

Disposal Options

There are a number of other research activities in progress concerning alternative disposal options. Two of these originate in Russia, where a number of graphite stacks contain high levels of contamination from fuel failures. One such process involves the vitrification of graphite waste, admixing the graphite with titanium dioxide and aluminium in a process akin to the 'Thermit' process. Development of this continues at The University of Sheffield (Dmitriev *et al.*, 2008) in the UK. The second utilises a molten-salt technology for graphite oxidation. This particular technology has been developed with particular reference to treatment of graphite heavily contaminated with "fuel spills" and involves air oxidation of the graphite and conversion of the actinides to oxides followed by vitrification of the resulting salt waste. The advantages claimed for this process include:

- Maximum capability for retaining radio-nuclides and heavy metals (up to 99.9% and higher);
- Lower operating temperature (750-900°C) as compared to combustion of graphite in flame furnace;
- Reprocessing of graphite without its preliminary crushing;
- Satisfactory dry cleaning of gases without scrubbers;
- Neutralization of acid gases; *and*
- Lower flow rates of flue gases as compared to combustion in flame furnace.

Separately, utilisation of microbial isotope separation is under serious consideration in South Africa (Dunzik-Gougar, 2008). This possibility builds upon a patented process of biofractionation which is applicable to uranium enrichment, and finds in initial investigations that the selected microbes selectively ingest [14]C.

Finally, a process of plasma-based cracking and re-oxidation of ^{14}C-bearing gases has been investigated theoretically in Japan (Mori *et al.*, 2006) in support of the dismantling of the Tokai 1 reactor, apparently in response to concerns about the impracticability of cryogenic separation (although incineration has previously been mooted for fuel sleeves and reflector blocks, so the situation regarding this reactor is confusing). These options are included here in brief because it is extremely important to remain open-minded about the options available for dealing with irradiated graphite. An optimal solution for one plant will not necessarily apply to another. However, the present authors believe that the most productive route for graphite treatment remains that illustrated in Figures 1 and 2.

RISK ANALYSIS

A feature of the debate relating to rad-waste disposal in general and to graphite disposal in particular is the perception of attendant risk. Much has been made of the 'seriousness' of radioactive waste disposal, where authorities appear to be working to risk levels far below those encountered in everyday life. Two recent publications attempt to address this issue (Neighbour *et al.*, 2000 and Neighbour and McGuire, 2008) whilst the demands of regulators and waste authorities to achieve levels way below background (and below the permissible emissions of extant nuclear installations by factors of up to 3000 in daily release rates for ^{14}C, for example) are challenged in a recent review of the graphite leaching data (Wickham, 2008). This cannot be fully debated here, but it is clear that all potential processes for graphite disposal need to be demonstrably safe and to work within national and international agreements on emissions until such time as sufficient concrete evidence exists to challenge established opinion on the 'true' risks represented by radioactive graphite. Williams (1998) admirably places the true risk of 'living in a nuclear age' in its correct, fact-based, context.

CONCLUSIONS

It is considered that a process of removal of graphite from reactor vessels which is conveniently described as 'nibble and vacuum' offers a very good alternative to direct burial, on the grounds of cost saving and reduction of waste volume, addresses 'waste hierarchy' issues, and offers scope for risk reduction through isotope dilution. Equally, isotope concentration can offer commercial advantages for isotope production, particularly of ^{14}C, and the method also allows for the re-incorporation of carbon from reactors into new graphite and carbon products for the nuclear industry. International investigations are continuing in the expectation that these ideas can be refined into a commercial and viable industrial process to facilitate the timely removal of graphite stacks in the decommissioning process.

References

Best, J. V., Wickham, A. J. and Wood C. J. (1976). Inhibition of moderator graphite corrosion in CEGB Magnox reactors – part 2: hydrogen injections into Wylfa Reactor 1 coolant gas. *J. Brit. Nucl. Energy Soc.,* **15**, 325-331.

Bradbury, D. and Mason B. (2008). Program on technology innovation: graphite waste separation. EPRI Palo Alto CA, 1016098.

Dmitriev, S. A., Karlina, O. K., Klimov, V. L., Pavlova, G., Yu. and Ojovan, M. I. (2008). Thermodynamic modelling of an irradiated reactor graphite thermochemical treatment process. *Solutions For Graphite Waste: A Contribution To The Accelerated Decommissioning Of Graphite Moderated Nuclear Reactors.* IAEA-TECDOC-1647 (*in press*).

Dunzik-Gougar, M. L., Chirwa, E. E. C., Chabalala, S., Pete, G. A., Bapela, I. I., Makgato, S., Kuczinski, L., van Ravenswaay, F. and Slabber, J. (2008). Microbial treatment of irradiated graphite for separation of radioisotope ^{14}C from bulk graphite ^{12}C. 7th EPRI International Decommissioning Meeting, Lyon, October 2008. *Proceedings*, 313.

Fachinger, J. (2008). Decontamination of nuclear graphite by thermal methods. *Solutions For Graphite Waste: A Contribution To The Accelerated Decommissioning Of Graphite Moderated Nuclear Reactors.* IAEA-TECDOC-1647 (*in press*).

Fisher, M. (1998). Fort St. Vrain decommissioning project. *Technologies for Gas-Cooled Reactor Decommissioning, Fuel Storage and Waste Disposal,* IAEA-TECDOC-1043, 123-131.

Marsden, B. J., Hopkinson, K. L. and Wickham, A. J. (2002). The chemical form of Carbon -14 within graphite. Serco Assurance Report SA/RJCB/RD03612001/R01 Issue 4 (*available for download from the website of the UK Nuclear Decommissioning Authority*).

Marsden, B. J. and Wickham, A. J. (1998). Graphite disposal options: a comparison of the approaches proposed by UK and Russian reactor operators. Proc. 'Decon '98, IMechE. 145-153.

Mori, S., Sakurai, M. and Suzuki, M. (2006). The recovery of Carbon-14 from the graphite moderator of a dismantled gas-cooled reactor through plasma-chemical reactions in CO glow discharge. *J. Nuclear Science and Technology,* **43**, 432-436.

Nair, S. (1983). A model for global dispersion of ^{14}C released to the atmosphere as CO_2. *Journal of the Society for Radiological Protection,* **3**, 16-22.

Neighbour, G. B. and McGuire, M. (2008). Aspects of graphite disposal and the relationship to risk: a socio-technical problem. *Solutions For Graphite Waste: A Contribution To The Accelerated Decommissioning Of Graphite Moderated Nuclear Reactors.* IAEA-TECDOC-1647 (*in press*).

Neighbour, G. B., Wickham, A. J. and Hacker, P. J. (2000). Determining the future for irradiated graphite disposal. *Nuclear Energy,* **39**, 179-186.

Ozanam, O. (2008). Current status and future objectives for graphite and radium-bearing waste disposal studies in France. *Solutions For Graphite Waste: A Contribution To The Accelerated Decommissioning Of Graphite Moderated Nuclear Reactors.* IAEA-TECDOC-1647 (*in press*).

Oziashvili, E. D. and Egiazarov, A. S. (1989). The separation of stable isotopes of carbon. *Russian Chemical Reviews,* **58**, 325-336.

Poncet, B. (2008). Data assimilation and dismantling: a methodology customised for radioactive inventory assessment purposes for activated metal, concrete or graphite. 7th EPRI International Decommissioning Meeting, Lyon, October 2008. *Proceedings*, 117.

Rahmani, L. (2008). Geological disposal of graphite waste as an aqueous foam. 7th EPRI International Decommissioning Meeting, Lyon, October 2008. *Proceedings*, 315-320.

White, I. F., Smith, G. M., Saunders, L. J., Kaye, C. J., Martin, T. J., Clarke, G. H. and Wakerley, M. W. (1984). Assessment of management modes for graphite from reactor decommissioning. Commission of the European Communities Report EUR 9232, Luxembourg.

Wickham, A. J. (2008). Graphite leaching: a review of international aqueous data with particular reference to the decommissioning of graphite-moderated reactors. EPRI, Palo Alto CA, 1016672.

Williams, D. R. (1998). What is safe? – the risks of living in a nuclear age. *The Royal Society of Chemistry, Cambridge.* ISBN 0-85404-569-4.

Should Subjective Risk be taken into Account in the Design Process for Graphite Disposal?

Michael A. McGuire, Gareth B. Neighbour[14] and Robert Price
Dept. of Engineering, University of Hull, Cottingham Road, Hull, UK. HU6 7RX
Email: g.b.neighbour@hull.ac.uk

Abstract

In terms of volume, irradiated graphite is by far the greatest contributor of low and intermediate level waste, and while deep storage may be the best option for high level waste, other solutions may need to be found for the larger volumes of less contaminated waste. This paper explores the relevance of the design process to potential solutions for irradiated graphite disposal. In particular, the paper seeks to evaluate the importance of subjective risk in the context of classical design theory in identifying appropriate solutions that satisfy the declared functional requirements. Furthermore, this paper recognises the different influences and constraints within the design process and suggests subjective risk is a key factor, and presents quantitative data that allow the different attributes of subjective risk to be incorporated into the decision process. A mathematical description is presented which relates objective risk to subjective risk and provides a conceptual framework for future studies. In essence, an argument is presented which suggests a decision-making process that does not include consideration of subjective risk may result in an inappropriate solution that does not meet with general acceptance.

Keywords

Subjective risk, Objective risk, Design process

INTRODUCTION

"What we anticipate seldom occurs; what we least expect generally happens."
- Benjamin Disraeli (1804 – 1881).

Much of the opposition to proposals for dealing with the radioactive waste arising from the decommissioning of nuclear power stations stems from irrational public concern; one might even call it hysteria that pervades the current debate. This irrationality stems from a wide range of qualitative factors that influence serious public consideration of the available options based on logical argument. Try as it may, and despite current preoccupation with the effects of climate change, the industry has largely failed to convince a sceptical public of the obvious advantages of nuclear power over other significant energy sources. Paramount among the arguments used against is the issue of nuclear waste, viewed as an undesirable legacy for future generations. In terms of volume, the graphite core is by far the greatest contributor of low and intermediate level waste, and while deep storage may be the best option for high level waste, other solutions need to be found for the large volumes of less contaminated waste that will be generated during the decommissioning programme (Nirex, 2005). The authors suggest that, in addition to standard arguments, the safety case for these categories of waste must address the more qualitative area of public perception of the dangers. In short, uninformed public perception of the risks is a hindrance to progress, and it is therefore important that a methodology for quantifying subjective risk be developed to take account of irrational fears stemming from ignorance, prejudice and emotion. Importantly, a decision-

---

[14] Author to whom any correspondence should be addressed.

making process that does not include consideration of subjective risk may result in an inappropriate solution that does not meet with general acceptance. Here, the authors argue that the problem of graphite disposal, in relation to general nuclear waste policies, has a socio-technical dimension related to subjective risk.

THE NATURE OF RISK

Conventional wisdom defines objective risk (R_O) as the product of the frequency of occurrence of a potentially hazardous event and the magnitude of the consequences. In a nuclear context, the consequences are measured in terms of radiation dose per annum per head of the population located at different distances from the source. In everyday life, the consequences of a hazardous event are usually measured in terms of death or serious injury, or in terms of the magnitude of an associated insurance claim or compensation award. Whether consciously or subconsciously, deciding whether a risk is worthwhile (acceptable) is an essential part of human existence, and thus the tolerability of risk in the public's mind is governed by a number of emotional responses, which are related to everyday experience. In this case, the assessment of risk is purely subjective, but no less real. Thus, there is a tendency for the layperson to underestimate the risks associated with activities where they feel in control, like driving a car, and overestimate those where control lies with someone else, like flying with a commercial airline. The social consequences of the former are borne out in the accident statistics. The extent to which an activity is voluntary, rather than imposed, is an important factor here, and, generally speaking, there is zero tolerance of a significant threat or hazard that is not voluntary (Starr, 1968). Broadly, the perceived or subjective risk (R_S) of a major hazard equates to a combination of the nature of the threat and feelings of outrage, disgust or dread, such as in the case of landmines, nerve gas or nuclear war. In the case of less tangible or immediate threats, other emotions play an important part, such as familiarity, trust and moral stance. Consequently, there is a range of attributes that contribute to such perceptions of risk, which may be summarised as follows:

- Naturalness (whether natural, artificial, industrial or customary)
- Familiarity (related to personal experience of the familiar or the exotic)
- Memorability (extent of past memories, particularly of bad experiences)
- Chronic state or catastrophic event (*i.e.* continuous or episodic)
- Knowable or unknowable (the ability to imagine the nature of the unknown)
- Tolerability (related to feelings of outrage, disgust or dread)
- Control (being controlled by oneself as against controlled by others)
- Morality (morally irrelevant as against morally relevant)
- Trustworthiness (*e.g.* institutions, agencies, industries, commercial brands)
- Benefit and reward (whether the outcome is beneficial or harmful)
- Fairness (to the individual or within society)

An essential element of risk is that it is usually balanced against a potential benefit. This is particularly true of the business world, and to technological situations involving investment decisions. It is important to distinguish between risk and uncertainty; objective risk is measurable, but there can be uncertainties in the probability of an activity or event and the predicted outcome. In addition, several probabilities may contribute to the overall probability of an event that leads to an undesirable outcome: the probability of the threat or hazard occurring (despite prevention measures), the probability of a vulnerability (for instance in protection systems) and the probability of the extent of resulting damage. Furthermore, several categories of risk can be identified: to the individual, to workers (occupational), to society (public at large or specific groups), economic (property and financial) and environmental, and these govern attitudes to risk and how risk is treated in relation to hazardous events. A hazard is usually defined as potential for danger, and can arise from causes that are natural in origin (*e.g.* earthquake), technological (*i.e.* man-made), social and moral (*e.g.* war, famine) or as a result of lifestyle choices (*e.g.* extreme sports). In addition, a hazard may exist in one of several states, including active, dormant, potential or mitigated (*i.e.* steps have been taken to alleviate the consequences should they arise).

METHODOLOGY

Price (2008) set out to measure public attitudes and responses to a range of events, activities, technological advances, experiences and practices in a systematic way, in order to lay the foundation for a rational approach to a possible quantification of subjective risk. To this end, a comprehensive survey was performed across a small, but representative, sample of the local population, including all age-groups, using a specially designed questionnaire. Participants (102) were asked to score, from -5 to $+5$, a wide range of potential hazards (64) against each of the attributes listed above, with the exception of fairness. The weighted scores for each activity, based on the number of responses ascribed to each value, were then plotted between pairs of attributes to identify the best correlations. Of particular interest were the responses the participants gave when asked at the outset to rank the ten attributes in order of importance to them. The two most significant attributes involved in subjective risk are the degree of perceived benefit or harm ($\sim 23\%$), and trustworthiness ($\sim 18\%$). This is particularly relevant to the subsequent discussion. The pairing of the results was also revealing. As expected, trustworthiness against benefit/harmfulness showed a one-to-one relationship with a close correlation. The combination with the highest correlation coefficient, and close to a one-to-one relationship, was found to be trustworthiness versus tolerability, Figure 1. This confirms that, in the public's mind, risk is more tolerable if the agency giving rise to the threat or hazard can be trusted to minimise the risk. This judgement is clearly a matter of direct personal experience, prejudice or hearsay, and it is unsurprising that the manner in which events are portrayed in the media can have a significant influence on such a judgement (HSE, 2001a). The MMR vaccine controversy is a good example. Tolerability and trustworthiness were also strongly linked to 'knowableness', although this attribute was rated lowest at the outset (2%). Clearly, fear of the unknown has a strong influence on perception of risk, and is closely related to familiarity which scored more highly. In view of the close relationship between several of the attributes, it is convenient to group some of them together, forming three main groups, as follows:

(A) Reward Attributes (R_e): Benefit/harmfulness

(B) Tolerance Attributes (T_o): Tolerability, morality, chronic/catastrophic, memorability, and naturalness

(C) Trust Attributes (T_r): Trustworthiness, control, knowableness, familiarity

Thus, in this discussion of subjective risk, three 'variables' can be defined: Reward (R_e), Tolerance (T_o) and Trust (T_r), the principal components of which are the key attributes of Benefit/Harmfulness, Tolerability and Trustworthiness.

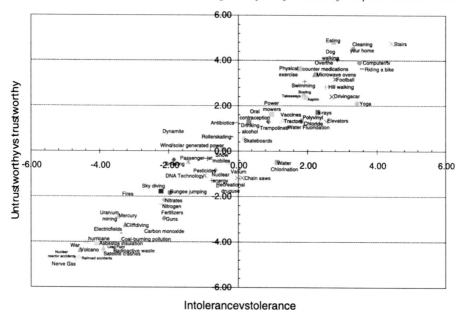

Intolerancevstolerance

FIGURE 1: Correlation between survey scores for trustworthiness and tolerability.

The extent to which the remaining variables contribute to the evaluation of R_e, T_o and T_r is discussed further below. In the authors' view, the relationship between these variables is the key to understanding public perceptions of risk, and how they might be quantified and applied to the topic of graphite waste. Formally, objective risk, R_o, as a consequence per unit time, might be defined as:

$$R_o = [\text{frequency (events/year)}] \times [\text{consequences (deaths or £ per event)}] \qquad [1]$$

Further, subjective risk, R_s, can be expressed in terms of R_o such that:

$$R_s = R_o \times F_s \qquad [2]$$

where F_s is a factor reflecting the extent of subjective bias. It can be seen from the above that F_s is the ratio of subjective to objective risk, *i.e.* R_s / R_o. Intuitively, F_s must be capable of assuming a wide range of values, greater than or less than unity, fully to reflect the range of views revealed by the questionnaire. Support for this model is provided by anecdotal evidence. It has been suggested by Starr (1968), for instance, that car-drivers perceive the risk of a motoring accident to be up to 1000 times lower than it actually is, making F_s for this activity, according to the above definition, equal to 10^{-3}. By the same token, other voluntary activities are rated as more dangerous than they really are, such as rail travel. On the basis that the public can underestimate or overestimate the apparent dangers by orders of magnitude, it is reasonable to ascribe a power function to the value of F_s. Thus, the definition of F_s can be rewritten as:

$$F_s = (R_s/R_o) = 10^{-(nS)} \qquad [3]$$

where S is the Subjectivity Index, being a function of R_e, T_o and T_r, and n is a scaling factor. The Subjectivity Index can further be expressed as:

$$S = f(R_e, T_o, T_r) = -\frac{1}{n}\log_{10}(F_s)$$

[4]

Thus, using the above example, nS = +3.0 for car driving. Together, the terms n and S constitute a Subjectivity Exponent. Thus, S represents the extent to which quantitative measures of objective risk must be adjusted to account for emotional bias whilst n takes account of the arbitrary nature of the scoring system used in the survey to identify attitudes to risk. The true value of n can only be derived by 'calibration' comparing subjective evaluations of risk, *e.g.* car driving example above, with actual statistics for objective risk.

In considering the relative contributions that the various attributes might make to the overall value of S, it is useful to consider some typical values of scores derived from the Hull Survey, as summarised for a representative sample of activities in Table 1, where the principal contributors (or key attributes) to each of the Groups A, B and C defined above are cited. For the car driving example, the survey returned figures of +2.74 for R_e, +2.42 for T_o, and +2.70 for T_r, on a range from +5 (Beneficial / Tolerable / Trustworthy) to –5 (Harmful / Intolerable / Untrustworthy). If these values are combined in a simple linear fashion, *i.e.* the average, the aggregated value for these three attributes becomes +2.62. Ignoring, for the time being, the other contributing attributes, this implies a value for n of 3.0/2.6 = 1.15. The Hull Survey included several nuclear-related activities / events in its scope with scores of -4.35 for harmfulness for a nuclear accident, -3.61 for nuclear waste and -0.31 for nuclear energy. A nuclear accident scored -4.73 for Tolerability, and -4.71 for Trustworthiness. Similarly, nuclear waste scored –4.34 and –3.60, respectively. It can be seen that these latter values are only marginally better than those for natural disasters, *e.g.* a hurricane or a volcanic eruption, and man-made threats such as war and nerve gas, illustrating the magnitude of the task involved in convincing the public that nuclear waste can be dealt with safely. Nuclear energy faired rather better against these attributes, scoring –1.30 for Tolerability and –0.75 for Trustworthiness. In order to incorporate the contributions from the other attributes in each group, some form of additional weighting is required that reflects the relative importance of each attribute against the other attributes. There are three possible candidates that could be considered as a basis for a weighting scheme: the initial ranking of the attributes by the participants in the Survey, quoted above, the quantitative results from the least squares correlations performed between the scores (see Figure 1), or a new set of factors derived separately using the Binary Dominance Matrix (BDM) method (see Hurst, 1999). The results of several correlations, in terms of the slope of the linear relationship describing the fit, and the degree of fit indicated by the correlation coefficient, R^2, relevant to the present discussion, are summarised in Table 2. Only the relevant correlations are shown for clarity.

Values in ordinary type are the results for relevant correlations within a group; values in bold type are for correlations for Tolerability and Trustworthiness with Benefit/Harmfulness. Most of these results show close correlations, with approximate one-to-one relationships. A notable exception is Naturalness, which was overall, rather poor as a descriptor. It is apparent that interpretations varied amongst participants, which has given rise to a wide scatter in the results, and consequently, a poor correlation with other attributes. Having considered the relative merits of these alternatives, the authors have opted for weighting factors derived using the BDM method. It was recognised that two sets of weighting factors are required, depending upon whether the scores being weighted are positive or negative. A positive score in the Survey is interpreted as indicating an *optimistic view* of a particular activity and a negative score a *pessimistic view*. This notion is discussed further below. The weighting factors obtained, W_o and W_p, to be applied to optimistic (positive) scores and to pessimistic (negative) scores, respectively, are summarised in Table 3:

TABLE 1: Survey results for a selection of potential hazards with differing threats for benefit / harmfulness, tolerability and trustworthiness.

Activity/Event	Key Attribute Scores from Survey Rated −5 to +5			
	Benefit/Harm	Tolerability	Trustworthiness	Average
Stairs	+3.15	+4.75	+4.43	+4.11
Football	+2.92	+3.18	+2.78	+2.96
Car Driving	+2.74	+2.42	+2.70	+2.62
Vaccines	+0.43	+1.36	+1.33	+1.04
Skateboarding	+0.98	+0.55	+0.17	+0.57
Alcohol	-0.08	+1.28	+0.31	+0.05
DNA Tech	-0.27	-1.09	-0.95	-0.77
Nuclear Energy	-0.31	-1.30	-0.75	-0.79
Passenger Jet	-0.66	-0.49	-1.43	-0.86
Dynamite	-2.31	+0.75	-1.65	-1.07
Bungee Jump	-1.15	-1.84	-1.97	-1.65
Smoking	-4.24	-0.39	-1.85	-2.16
Cliff Diving	-2.40	-3.25	-3.25	-2.97
Nuclear Waste	-3.61	-4.34	-3.60	-3.85
War	-3.81	-4.17	-4.62	-4.20
Hurricane	-4.47	-4.01	-4.38	-4.29
Volcano	-4.37	-4.36	-4.57	-4.43
Nucl Accident	-4.35	-4.73	-4.71	-4.60
Nerve Gas	-4.60	-4.91	-4.88	-4.80

TABLE 2: Correlations between attributes.

	Attribute Group	Attribute	Ranking at Outset (%)	Slope	Correlation Coefficient, R^2
A	REWARD (Re)	**Benefit/Harmfulness**	23	-	-
B	TOLERANCE (To)	**Tolerability**	14	**0.83**	**0.84**
		Morality	11	0.83	0.64
		Memorability	5	0.90	0.71
		Chronic/Catastrophic	3	0.91	0.65
		Naturalness	3	0.39	0.17
C	TRUST (Tr)	**Trustworthiness**	18	**0.91**	**0.85**
		Control	15	0.74	0.39
		Familiarity	8	0.80	0.56
		Knowableness	2	0.92	0.72

TABLE 3: Weighting factors derived using Binary Dominance Matrix method.

ATTRIBUTE	WEIGHTING FACTORS	
	Wp (pessimistic)	Wo (optimistic)
Naturalness	0.089	0.111
Familiarity	0.067	0.123
Memorability	0.178	0.022
Chronic/Catastrophic	0.133	0.067
Knowableness	0.133	0.067
Tolerability	0.089	0.111
Control	0.067	0.123
Morality	0.200	0.000
Trustworthiness	0.000	0.200
Benefit/Harmfulness	0.044	0.156

The Weighting Factors from Table 3 have been applied to the individual Survey scores, and the results summed for the attributes within each group, Table 4. Moreover, S can then be more definitively expressed as follows:

$$S = R_e + T_o + T_r \qquad [5]$$

It then follows that the Subjectivity Exponent, nS, is obtained by scaling all S values to car driving (*i.e.*, $nS = +3.00$) where $n = (3.00/1.80) = 1.66$. If Table 4 results are compared with those in Table 1, it can be seen that the order of the activities and events, listed according to the values quoted in the final column of each Table, changes slightly, but not substantially. Table 4 shows that combining the Survey scores, using the BDM weighting method, re-balances the results, such that natural disasters, *e.g.* volcanic eruptions and hurricanes, although still rated pessimistically (S is still negative), are now separated from man-made threats such as war, nuclear accidents and nerve gas, a more rational result. Values of the Subjectivity Exponent, nS, for the full range of potential hazards employed in the Hull Survey are shown in Figure 2. In order to place these results in context, values of Subjective Risk have been derived for several representative examples, using the methodology presented above, and compared with corresponding values of Objective Risk, Table 5 and Figure 3. The objective risk statistics for being struck by lightning and dying from all causes are included for completeness. Overall, the results may be interpreted as follows. Values of the Subjectivity Exponent, nS, close to zero (*i.e.* $F_s = 1.00$) indicate that subjective risk coincides with objective risk, and suggests that the public have a 'realistic' view of the potential hazard. Equality between objective and subjective risk is indicated by the solid line in Figure 3 ($nS = 0$). Thus, the Hull methodology gives $nS = +0.199$ for smoking, and $+1.026$ for agricultural accidents involving tractors; hence, these lie close to the $nS = 0$ line in Figure 3. By contrast, according to the methodology, flying as a passenger is perceived to be about 200 times more hazardous than it is in reality, with $nS = -2.33$. This point therefore lies closer to the dashed line representing $nS = -3.0$. The example of car driving ($nS = +3.00$) has been discussed above, and is used as the basis for calculating n. Put more simply, positive values of nS reflect an optimistic view of the potential hazard and negative values a pessimistic view. Broadly, the range of nS values can be conveniently divided into four bands, Figure 2, representing a range of views from extreme optimism to extreme pessimism: $nS > +3.0$ (very optimistic view); $nS = 0.0$ to $+3.0$ (optimistic view); $nS = 0.0$ to -3.0 (pessimistic view); $nS < -3.0$ (very pessimistic view).

TABLE 4: Evaluation of Subjectivity Risk Index, S, and Subjectivity Exponent, nS, for a selection of potential hazards (compare with Table 1).

Activity/Event	Weighted Scores for Attribute Groups using Table 3 Values				
	REWARD (Re)	TOLER-ANCE (To)	TRUST (Tr)	S (Sum of Re, To and Tr)	Subjectivity Exponent, nS *
Stairs	+0.49	+1.15	+2.22	+3.87	+6.43
Football	+0.46	+0.88	+1.55	+2.89	+4.80
Car Driving	+0.43	-0.02	+1.40	+1.80	+3.00
Alcohol	-0.00	+0.17	+1.28	+1.45	+2.41
Skateboard	+0.15	+0.27	+0.82	+1.24	+2.06
Vaccines	+0.07	+0.01	+0.15	+0.22	+0.37
Smoking	-0.19	-0.44	+0.74	+0.12	+0.20
Passenger Jet	-0.03	-1.03	-0.34	-1.40	-2.33
Bungee Jump	-0.05	-1.60	-0.01	-1.67	-2.77
Dynamite	-0.10	-1.48	-0.62	-2.20	-3.65
Volcano	-0.19	-0.94	-1.07	-2.21	-3.68
Hurricane	-0.20	-1.02	-1.06	-2.27	-3.78
Nuclear Energy	-0.01	-2.00	-0.34	-2.34	-3.90
DNA Technology	-0.01	-1.58	-0.80	-2.39	-3.98
Cliff Diving	-0.11	-2.54	+0.12	-2.53	-4.21
Nuclear Waste	-0.16	-2.09	-0.64	-2.89	-4.81
War	-0.17	-3.16	-0.99	-4.31	-7.17
Nuclear Accident	-0.19	-3.23	-1.02	-4.44	-7.39
Nerve Gas	-0.20	-3.32	-1.10	-4.62	-7.69

* Scaled to Car Driving = 3.00 (n = 1.66)

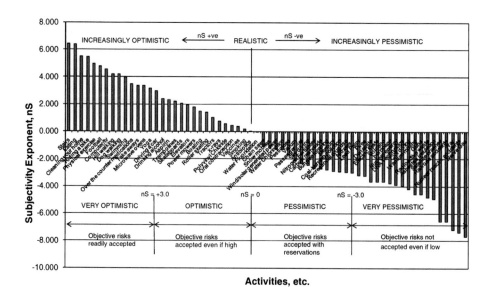

FIGURE 2: Subjectivity exponents for the full range of potential hazards.

TABLE 5: Comparison between estimates of objective risk and subjective risk calculated using the Hull methodology.

Point No.	Activity/hazard	R_O (Deaths per annum)	Reference	nS	Fs	R_S (=Ro.Fs) per annum (derived)
1	Asbestos-related disease	3.38×10^{-5}	HSE Statistics 2008	-4.937	8.650×10^{4}	2.92
2	Radioactive waste	1.00×10^{-6}	HSE (2001b), HSE (2006)	-4.812	6.486×10^{4}	6.49×10^{-2}
3	Rail accidents	6.76×10^{-9}	HSE (2001b),	-6.584	3.837×10^{6}	2.59×10^{-3}
4	Smoking-related disease (estimate)	2.00×10^{-3}	National Statistics (2006, 2008), HSE (1999).	+0.199	0.6324	1.26×10^{-3}
5	UK Nuclear Industry	2.50×10^{-8}	Williams (1998)	-3.898	1.265×10^{4}	3.16×10^{-4}
6	Flying as a passenger	1.00×10^{-7}	HSE (2001b),	-2.326	2.118×10^{2}	2.12×10^{-5}
7	Agricultural accidents (including tractors)	8.10×10^{-5}	HSE Statistics 2008	+1.026	9.419×10^{-2}	7.63×10^{-6}
8	Driving a car	7.16×10^{-5}	Starr (1968)	+3.000	1.000×10^{-3}	7.16×10^{-8}
9	Swimming/drowning	8.00×10^{-6}	ROSPA (2001)	+3.986	1.033×10^{-4}	8.26×10^{-10}
10	Football	4.00×10^{-5}	Williams (1998)	+4.801	1.581×10^{-5}	6.32×10^{-10}
11	Struck by lightning	6.67×10^{-8}	HSE (2001b),	-	-	-
12	Risk of death from all causes	1.136×10^{-2}	HSE (2001b),	-	-	-

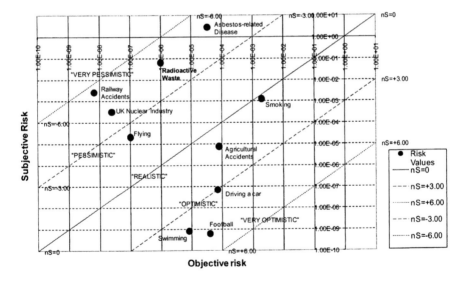

FIGURE 3: Values of subjective risk calculated using the Hull methodology plotted against actual values of objective risk for a selection of potential hazards.

An optimistic view of the potential hazard equates to a tendency to accept objective risks 'even if high', whereas a pessimistic view equates to objective risks being 'accepted with reservations'. At the more extreme ends of the range, excessive optimism means that objective risks are probably readily accepted whatever they are, whereas extreme pessimism means that objective risks are not accepted even if they are demonstrably low. By way of illustration, railway accidents, nuclear energy and flying are rated as pessimistic or very pessimistic, even though the objective risks are very low. On the other hand, commonplace pursuits like swimming and playing football are, unsurprisingly, rated as 'very optimistic' (nS > 3.0), whereas asbestos-related disease is rated 'very pessimistic' (nS < -3.0), even though these have very similar values of objective risk. While no formal relationship exists between these calculated values and actual values of risk, they provide a strong indication of public attitudes to potential hazards. Furthermore, the model provides a consistent basis on which to monitor changes in public opinion over time. This is particularly important where public acceptance of potentially controversial projects is the ultimate objective (*e.g.* Heathrow 3rd runway, onshore wind farms, nuclear waste repository).

APPLICATION OF THE MODEL TO DISPOSAL OF GRAPHITE WASTE

White *et al.* (1984) assessed a range of options for the management of waste graphite from the Magnox reactor decommissioning programme, including disposal both on land and at sea. However, following the OSPAR Convention (Gray, 1998), disposal at sea is no longer considered acceptable. Options for disposal on land included a deep geologic repository, shallow burial and incineration. A key consideration in the choice and acceptability of management options was the potential radiological impact of each option on man and the environment. Detailed radiological assessments yielded the following values for potential individual doses (Sv/y):

Inland geologic disposal	8.3×10^{-10} to 1.3×10^{-6}
Coastal geologic disposal	8.3×10^{-10} to 1.1×10^{-4}
Shallow land burial	4.3×10^{-7} to 1.7×10^{-2}
Incineration: atmosphere	6.7×10^{-4}
Incineration: ash burial	4.8×10^{-6} to 1.9×10^{-1}

To place these values in context, HSE (2006) states that the current legal limit to individual members of the public is 1 mSv/y (Basic Safety Level), although the Basic Safety Objective (BSO) is 0.02 mSv/y. HSE (2001b) states that, in terms of cancer-induced deaths per annum, this is equivalent to an overall (objective) risk of 10^{-6} p.a. This value is included in Table 5, and compared with the value calculated for subjective risk associated with radioactive waste, Figure 3. A value is also included for the nuclear industry as a whole. Both these receive a 'very pessimistic' rating, despite low values of objective risk. This confirms the notion that, while the public might be relatively sympathetic to the concept of nuclear energy, they will not tolerate the waste it produces, at least, not under the current arrangements. Persuading the detractors that one is a necessary by-product of the other, and can be safely dealt with, is the key to ultimate acceptance. Clearly, in order to convince a sceptical public of the safety of the chosen option for disposal of graphite waste, subjective risk needs to be minimised. In the context of the Hull methodology, this means influencing public opinion such that the point in Figure 3 representing radioactive waste moves closer to the $nS = 0$ line, representing a more realistic view. How this might be achieved is of great importance to the future of nuclear energy in western economies. Encouragingly, issues of climate change and global warming have re-ignited the debate, and it is now being acknowledged, perhaps reluctantly, that nuclear power has a lot to offer in reducing carbon emissions. The issues surrounding the psychology of acceptance are intimately connected with subjective risk, and so it is necessary to target those attributes that contribute to Reward, Tolerance and Trust, Figure 4. Hence the decision-making process must counter perceptions of Harmfulness by instilling public Trust in the agencies responsible, whilst continuing to emphasise the positive Benefits of nuclear energy as part of a mix of new generating capacity. This can only be achieved by improving familiarity with the topic and by enhancing feelings of control through wider consultation and informed debate.

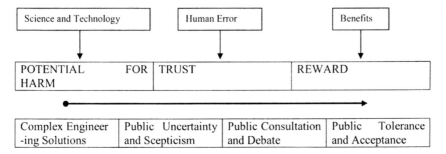

FIGURE 4: Illustrating progression towards public acceptance.

Neighbour and McGuire (2008), in a previous paper on this topic, whilst acknowledging that the circumstances are very complex, nevertheless pointed out that, to date, the decision-making process has been artificially constrained by a failure to employ a classical design process, especially in the use of conceptual design techniques. It was pointed out that finding a solution to the graphite waste problem was tantamount to a *design process*, in which a *design space* envelops all possible solutions to a design problem. In particular, it was shown that it was important first to identify "the customer" or end-user, and then to engage the customer, displaced or otherwise, fully in the process. In essence, the greater the number of ideas and concepts generated, the greater the probability the best solution will be found. The common practice of *optioneering* to identify the optimum engineering solution amongst a limited number of options, to satisfy the immediate requirements of a safety case, is an example where design constraints can be introduced too early in the process. This is because optioneering often bypasses the conceptual design stage, focussing on detailed design concepts prematurely. The design space must first be expanded using divergent thinking, followed by convergence to arrive at the optimum solution over several iterations. In complex situations, such as the graphite waste problem, there is a danger that traditional engineering approaches that are 'incremental' rather than 'disruptive' will not achieve the best solution. An essential feature of design theory is to identify Functional Requirements (FR) at the outset that are satisfied by a set of independent Design Parameters (DP), and which meet perceived needs. In the case of graphite waste, the authors reduced the functional requirements to: protect human health and the environment; solution to be sustainable *vis à vis* future generations; seek a high position in the hierarchy of waste (reuse/recycle); and, unmanned containment of radioactive species. These FRs, in turn, lead to the identification of other design criteria and attributes, and this is where the issue of subjective risk arises. Designs often fail because market and customer research (needs analysis) are inadequate, leading to a poor design specification. To address this shortcoming, a further FR should be added which addresses both objective and subjective risk. Most proposed solutions to the graphite waste problem to date have inevitably focussed on detailed technical issues aimed at minimising objective risk. While this must be an essential part of any safety case to satisfy the Safety Assessment Principles (HSE, 2006), the inclusion of this FR will ensure that public perceptions of the dangers of complex (and unfamiliar or unknowable) technologies will be taken into account, leading to earlier acceptance of future proposals.

It is worth commenting that the UK Government's White Paper (2008), "Managing Radioactive Waste Safely", published in June 2008, provides a framework for implementing geological disposal following the recommendations of the CoRWM Committee on Radioactive Waste Management (2006). CoRWM recognised that a successful outcome depends upon instilling public and stakeholder trust and confidence through an open and transparent process, and, to this end, invokes the concept of 'volunteerism'. Communities are to be invited to 'host' the proposed disposal facility and express a willingness to participate, participation being based on "the expectation that the well-being of the community will be enhanced". Recommendations include inviting host communities to participate in a flexible and staged decision-making process, a partnership approach and the right to withdraw at any stage. Clearly, this is an attempt to manage subjective risk, and is, therefore, fully consistent with the socio-technical approach proposed in this paper. In summary, this paper has added a mathematical framework to the earlier work by Neighbour and McGuire (2008) on the understanding of the role of subjective risk in the design process in managing radioactive waste.

CONCLUSIONS

A methodology has been described that quantifies subjective risk in terms of a range of attributes, based on the results from an attitude survey carried out by the University of Hull. The results have been combined, using suitable weighting factors, to yield a single parameter, the Subjectivity Index that can be used to factor values of objective risk to yield corresponding values of subjective risk. The methodology is extended to illustrate how it might be applied to the problem of radioactive waste, and, in particular, the options available for managing graphite waste from decommissioning programmes, by including subjective risk as a functional requirement in the design process. Finally, it is concluded that a decision-making process that does not include consideration of subjective risk may result in an inappropriate solution that does not meet with general acceptance. Further, this mathematical framework aids the decision making process in that it allows the mapping of the change of subjective risk over time.

References

Committee on Radioactive Waste Management (CoRWM 2006). Managing our radioactive waste safely: CoRWM's recommendations to Government. CoRWM Doc.700

Health and Safety Executive (1999). Proposal for an approved code of practice on passive smoking at work.

Health and Safety Executive (2001a). The impact of social amplification of risk on risk communication. Contract Research Report 332/2001, University of Surrey. ISBN 0-7176-1999-0.

Health and Safety Executive (2001b). Reducing risks, protecting people. ISBN 0-7176-2151-0.

Health and Safety Executive (2006). Safety assessment principles for nuclear facilities. Revision 1.

Health and Safety Executive (2008). HSE Statistics – Table MESO04: Number of Mesothelioma deaths and average annual rates per million by age and sex in three year periods, 1969-2005. http://www.hse.gov.uk/statistics/tables/meso04.htm

Hurst, K. (1999). Engineering design principles. Arnold, London. ISBN 0-340-59829-8.

National Statistics (2006). Statistics on smoking: England, 2006.

National Statistics Online (2008). http://www.statistics.gov.uk.

NIREX (2005). The 2004 UK Radioactive Waste Inventory CD-ROM. Version 1.

Neighbour, G. B. and McGuire, M. A. (2008). Aspects of graphite disposal and the relationship to risk: a socio-technical problem. In: Proceedings "Solutions for Graphite Waste: A Contribution to the Accelerated Decommissioning of Graphite-Moderated Nuclear Reactors". IAEA-TECDOC-1647 – In Press.

Price, R. (2008). The quantification of subjective risk and its relationship to product design. BSc Dissertation, Department of Engineering, University of Hull.

Gray, J. (1998). OSPAR Commission and Ministerial Meeting, 20-24 July 1998, Sintra, Lisbon. *J. Radiol. Prot.* **18** 306-310.

ROSPA (2001). Drownings in the UK. http://www.rospa.com/leisuresafety/water/statistics/2001statistics.htm

Starr, C. (1968). Social benefit versus technological risk. what is our society willing to pay for safety? *Science*, 165, 1232 – 1238.

UK Government White Paper (2008). Managing radioactive waste safely: a framework for implementing geological disposal.

Williams, D. R. (1998). What is safe? The risks of living in a nuclear age. The Royal Society of Chemistry.

White, I. F., Smith, G. M., Saunders, L. J., Kaye, C. J., Martin, T. J., Clarke, G. H. and Wakerley, M. W. (1985). Assessment of management modes for graphite from reactor decommissioning. European Commission Report EUR 9232, 1985.

The Location of Radioisotopes in British Experimental Pile Grade Zero Graphite Waste

Lorraine McDermott, Abbie N. Jones, Barry J. Marsden,
T. James Marrow and Anthony J. Wickham
The University of Manchester, Sackville Street, Manchester, M60 1QD.
Email: lorraine.mcdermott@postgrad.manchester.ac.uk

Abstract
The UK has approximately 90,000 tonnes of irradiated graphite waste accumulated since the 1940s from over 40 nuclear reactors (NDA, 2002 & DEFRA, 2005). In order to make an informed decision as to how to best deal with this waste, information on the activation and location of impurities contained within the graphite porous structure is required. In addition possible mechanisms that may lead to the release of these isotopes must also be well understood, not only to assess the possibility of release after disposal, but also to consider it may be possible to "clean" the graphite using thermal or chemical treatment thus significantly reducing the activity. The activities of isotopes contained within nuclear graphite may be theoretically calculated from the trace elemental impurities present within virgin graphite material and the cross sectional areas of these elements (Jones *et al.*, 2008; Hashemi, 2002; Forrest and Kopecky, 2007; and Podruzhina, 2004). This combined with reactor operational conditions provides background to the isotopic inventory currently accepted. However, other isotopes may arise from impurities trapped in the porous graphite during reactor operation. These activated impurities will need to be accounted. This paper presents microstructural and radiochemical techniques used to quantify the isotopic location and distribution within the graphite. These impurities have been characterised in terms of location and retention using high resolution techniques such as Scanning Electron Microscopy, Raman, micro X-ray Tomography and Energy Dispersive X-ray Spectroscopy.

Keywords
Nuclear graphite, Microstructural characterisation, Autoradiography, X-ray tomography

INTRODUCTION

The UK graphite moderated reactors will have produced somewhere in the region of 90,000 tonnes of irradiated nuclear graphite after operation ceases (DEFRA, 2005). In order to make informed decisions of how best to dispose of this large volume of waste it is necessary to understand the character of the irradiated graphite waste and the effectiveness of the various proposed decontamination and immobilisation treatments. The objective is to use microstructural and radiochemical techniques in order to quantify the isotopic location, distribution and chemical form within the irradiated graphite waste (Godbee *et al.*, 1969; Gray and Morgan, 1988; and Handy, 2006). This data will be used to compare with theoretically calculated isotopic inventory from trace elemental analysis. British Experimental Pile Zero (BEPO) was an experimental research reactor counterpart to the Windscale Pile plutonium production reactors. Low Power output of 6MW with operational lifetime of 20 years. BEPO used Canadian sourced graphite which was annealed once after Windscale piles accident 40 years since operation isotope concentrations in BEPO are principally tritium, Carbon-14, Cobalt-60 Europium-152 and Europium-154.

History of BEPO

British Experimental Pile Zero (BEPO) Energy Reactor was commissioned in 1948 and closed in 1968. BEPO was graphite moderated with horizontally lying channels, air cooled and initially fuelled with natural uranium metal in aluminium cans. Two grades of enriched uranium fuel were used at a later date within the reactor. BEPO was initially used for the production of plutonium, later this was transferred to the Windscale piles with BEPO primary function as a research reactor (Wickham, 2002) principally for isotope production and studying the irradiation behaviour of graphite. In addition BEPO was also used in studies for coolant compositions for the Magnox and AGR reactors.

History of BEPO Graphite

In 1975, a four inch core was trepanned from the BEPO core. The trepanned sample was drilled through the control face of the pile, steel shielding, concrete bio-shield and the 20 columns of graphite block (Wickham, 2002). Five samples of the core have been used for research purposes at the University of Manchester. The activity of the samples is related to their position in the core and is depicted in Figure 1. Sample 1 was located close to the outer surface of the channel and thus exhibits the least activity, in comparison sample 21 is located closest to the reactor core. This four inch diameter trepanned core was initially taken to provide radioisotope data on the graphite, steel and concrete, in addition BEPO was used to evaluate the Wigner energy, the trepanned core was retrieved using a diamond-tipped hole cutter using no coolant or lubricant. After retrieval the reactor core was resealed and to date remains this way at Harwell.

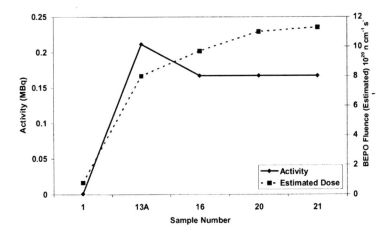

FIGURE 1: Estimated activity (MBq) and fluence of BEPO samples.

This study concentrates on developing methods to accurately determine the activity of the radioisotopes, principally ^{14}C and ^{3}H. The aim is to understand the nature and quantity of impurities in the initial microstructure and the effect on the microstructure of the graphite after neutron irradiation.

METHODS OF ANALYSIS

Autoradiography

Autoradiography provides a visual distribution pattern of radiation present in the surface of a sample, depending on isotope energy (Bauer, 1961). Autoradiography can therefore determine β and γ radiation present within nuclear graphite with energy above 0.018MeV. Weak β-emitting isotopes and α isotopes are stopped by the coating on the phosphorous storage film (recording medium). Autoradiography is a qualitative technique used to analyse high energy β and γ isotopes present within a nuclear graphite sample.

The phosphor storage screens comprise of fine crystals of $BaFBr:Eu^{+2}$ in an organic binder. Upon exposure to radiation the Eu^{+2} is excited to the oxidised Eu^{+3} state and the BaFBr is reduced to $BaFBr^{-}$. The phosphor storage screen releases this energy when exposed to light of an appropriate wavelength. Excited electrons fall to the ground state thus releasing energy in the form of blue light. The emitted light passes through a photomultiplier detector which converted to produce an electric current proportional to the activity in the sample.

The advantages of this technique include reduced exposure time compared to traditional autoradiography using X-ray film and increased sensitivity with a linear dynamic range of 1 to 100000 which allows both weak and strong energy isotopes to be analysed simultaneously.

X-ray Tomography

Computerised X-ray tomography is a non destructive technique that allows the three dimensional microstructure of irradiated graphite to be analysed. This technique involves sending an X-ray beam thorough the material and recording the transmitted beam using a charged coupled device camera as shown in Figure 2. The ratio of transmitted to incident photons is related to the integral of the absorption coefficient of the material along the path that the photons made through the material. Using an empirical law, the absorption coefficient can be related to the density, the atomic number and in certain cases, the energy (Berre *et al.*, 2006; Babout *et al.*, 2005; and, Babout *et al.* 2008). The resulting image is a projection of a volume in a given two dimensional plane. The method of acquiring a three dimensional image is to obtain radiographs whilst rotating the sample between 0° and 180°. An algorithm (filtered back-projection) is applied to reconstruct the volume from the two dimensional radiographs.

This CT data was used to determine porosity and density of the BEPO graphite. The smallest porosity or characteristics that can be identified however are limited to the resolution of the two dimensional radiographs which is determined by the sample size and CCD camera used.

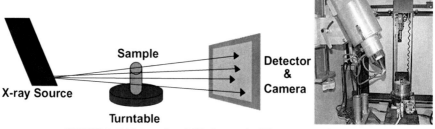

FIGURE 2: (A) Schematic and (B) photograph of X-ray tomography scanner.

EXPERIMENTAL PARAMETERS

Autoradiography

Amersham Typhoon 9410 has been used throughout this study. Sample size, or preferentially diameter is shown to have significant effect on the autoradiography results. Autoradiography analysis of graphite samples thicker than 1cm have proved difficult to analyse with the high energy radionuclide's saturating the film after only a 4 - 6 hours, making comparison between graphite samples that have acquired varying levels of dose or exposure inconclusive.

In total 60 samples where analysed. The sample size was 1mm in diameter and manually cut from using a saw. The autoradiography screen was wiped clean after each use using a light box and a lint free cloth. The samples were placed on the screen in a glove box and left for 20 hours. This length of time was determined as sufficient exposure time without saturating the film due to the activity present in the graphite and will vary for each type of sample being analysed.

Analysis of the film performed using the Amersham (wavelength of 633nm) and a pixel size of 50 microns. The average intensity, standard deviation, variance, minimum and maximum intensity and area are calculated. This allows the background to be deducted from analysis and documents any errors between results.

X-ray Tomography

Tomography images were collected using the X-TEK HMX CT instrument using the parameters outlined in Table 1. In addition, a rotation of 0.3° with 32 frames and a camera exposure time of 160s were also applied. A voxel size of 21.4 μm was resolved.

Irradiated graphite analysis was performed with the samples contained within a 'Harwell can', this ensured that no radioactive graphite dust would contaminate the tomography equipment. No filter was applied during the analysis as the can is made out of aluminium thus acting as a filter. A calibration of the can's thickness versus attenuation using highly orientated pyrolytic graphite (HOPG) as a reference standard inside the can is shown in Figure 3 where intensity shown to exhibit a linear relationship with thickness. The reconstruction of the two dimensional images was completed using Amira© software and are shown in Figure 6.

TABLE 1: Tomography experimental parameters.

Parameters	Settings
Target	Copper
Filter	None
Detector	Beryllium
Voltage	95 kV
Current	105 μA

EXPERIMENTAL RESULTS

Autoradiography

Autoradiography was carried out on all BEPO samples in order to gain a qualitative assessment of the distribution of radioisotopes within the irradiated BEPO material. Figure 4 shows the average intensity of the samples with increasing sample number *s* after an exposure time of 20 hours. Sample 1 exhibits the least intensity of the BEPO materials, which corresponds to the least active samples *s*. There is a non uniform distribution of radioisotopes which can be noted by the presence of high intensity regions contained within sample 16 this is also a high distribution of activity with both sample 20 and 21 this film is fully saturated and again shows a non uniform disruption of β and γ activity within the sample.

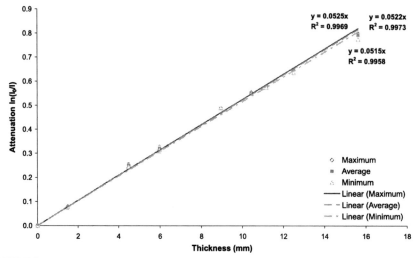

FIGURE 3: Calibration graph of thickness of HOPG sample inside a harwell can versus the attenuation intensity of the transmitted incident photons that have passed thorough the HOPG sample.

A)

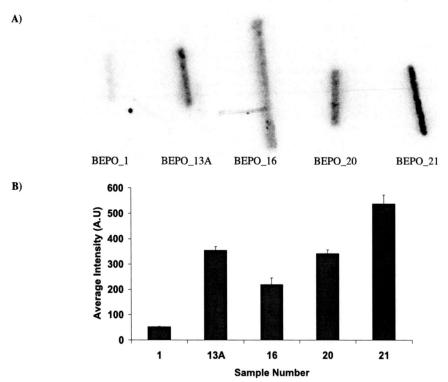

B)

FIGURE 5: (A)Autoradiograph showing increasing greyscale intensities with exposed BEPO samples, (B) graph showing the average intensity for BEPO samples.

X-ray Tomography

In each of the 2D slices spots of high attenuation are observed. These spots are of a material with a much higher density and therefore elemental number than that of graphite. The majority of these high attenuation spot are noted to be next to or contained within areas of porosity. Further work needs to be carried out using SEM-EDX to characterise these regions of interest.

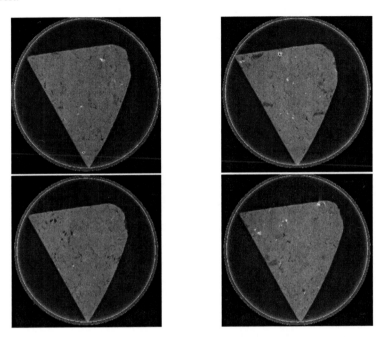

FIGURE 6: 2D tomography Radiographs of BEPO sample 16 graphite contained within a Harwall can.

Three dimensional volume reconstructions of BEPO sample 16 have been processed using Amira© software and are shown in Figure 7. The porosity distribution was calculated and is equivalent to 17.2% porosity. The blue regions display the internal porosity and the red is the high attenuation spots. It can be noted the is these high attenuations spots are contained within the porosity and are non-uniformly distributed.

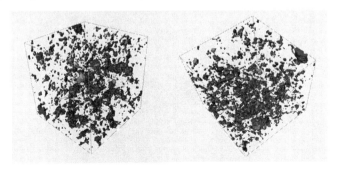

FIGURE 7: Three dimensional volume reconstructions of BEPO Graphite using Amira© software.

CONCLUSIONS

Autoradiography was successfully used to characterise the distribution of radioisotopes in BEPO graphite. Utilising thin 1mm thick samples of graphite allows for a reduction in gamma swapping effects and has enabled the investigation of anomalies within the radioisotopic distribution. The autoradiographs show there is a non-uniform distribution of radioactivity within the samples, this has also been highlighted using X-ray tomography, where high attenuation spots are seen and are attributed to a material with a molecular weight greater than that of graphite. In addition X-ray tomography has been used to analysis the shapes and distribution of porosity in irradiated BEPO.

Acknowledgements

The authors would like to acknowledge this work was carried out as part of the CARBOWASTE Programme: *Treatment and Disposal of Irradiated Graphite and Other Carbonaceous Waste,* Grant Agreement Number: FP7-211333 and partial funding from the NDA.

References

Bauer, E. (1961). Studies of the initial phases of oxidation of nuclear graphitein air at temperatures between 420 and 650 °C. *J. Chim. Phys. Rev. Gen. Colloides*; 58, 47-52.

Berre, C. *et al.* (2006). Numerical modelling of the effects of porosity changes on the mechanical properties of nuclear graphite. *Journal of Nuclear Materials*, 352, [1-3], 1-5.

Babout, L. *et al.* (2005). The effect of thermal oxidation on polycrystalline graphite studied by X-ray tomography. *Carbon*, 43, [4], 765-774.

Babout, L., *et al.* (2008). Three-dimensional characterization and thermal property modelling of thermally oxidized nuclear graphite. *Acta Materialia*, 56, [16], 4242-4254.

Defra (2005). Radioactive wastes in the UK: a summary of the 2005 inventory. Nirex.

Forrest, R. A. and Kopecky, J. (2007). The activation system EASY-2007. *Journal of Nuclear Materials*, 386-388, 878-881

Godbee, H. W., *et al.* (1969). Diffusion of radioisotopes through solid wastes. *Trans. Amer. Nucl. Soc.*, 12: 450-1. From 17th Conference on Remote Systems Technology, San Francisco, Calif. See CONF-691102.

Gray, W. J. and Morgan, W. C. (1988). Leaching of 14C and 36Cl from Hanford Reactor Graphite, P.N. Labs.

Handy, B. J. (2006). Experimental study of C-14 and tritium release from irradiated graphite spigot samples in alkaline solutions, Nirex, AMEC NNC: Didcot, 1-11.

Hashemi-Nezhad, S. R. et al. (2002). Monte Carlo calculations on transmutation of trans-uranic nuclear waste isotopes using spallation neutrons: difference of lead and graphite moderators. *Nuclear Instruments and Methods in Physics Research Section A: Accelerators, Spectrometers, Detectors and Associated Equipment*, 482, [1-2], 547-557.

Jones, A. N. *et al.* (2008). Microstructural characterisation of nuclear grade graphite. *Journal of Nuclear Materials*, 381, [1-2], 152-157.

NDA (2002). Radioactive wastes in the UK: a summary of the 2001 Inventory.

Podruzhina, T. (2004). Graphite as radioactive waste corrosion behaviour under final repository conditions and thermal treatment. In FZJ Report 2004: Jülich.

Wickham, A. J. (2002). BEPO graphite: compilation of data relevant to test of GIRM methodology. Department of Trade and Industry. p. 59.

Radioactive Gas Generation and Containment in ILW Repositories

Elina M. Kuitunen and Michael A. Hicks

School of Mechanical, Aerospace and Civil Engineering, The University of Manchester
Sackville Street, Manchester, M60 1QD, UK
Email: elina.kuitunen@postgrad.manchester.ac.uk

Abstract

Radioactive wastes placed in a repository can degrade by several mechanisms giving rise to gases. Although the bulk of the gas is expected to be hydrogen, gases labelled with radionuclides such as carbon-14 and tritium are also likely to form. The migration of these radionuclides from the repository can be retarded by the use of a barrier system, which includes the host rock, as well as engineered barriers such as waste packages and backfill. In the UK, graphite is a major source of C-14 and therefore understanding its behaviour in repository conditions is vital for safety assessments.

Keywords
Carbon-14, Gas generation from wastes, Multi-barrier system

INTRODUCTION

Radioactive wastes placed in the repository generate gases as they evolve. Although the bulk of the gas is expected to be hydrogen, gases labelled with radionuclides are also likely to form. This paper is a first stage towards identifying factors controlling repository gas generation, and subsequent migration and attenuation of gases in the near field, with a view to influencing future repository design.

WHAT RADIONUCLIDES ARE THERE?

Tritium
Tritium is widely distributed in radioactive wastes. It has a relatively short half-life of 12.35 years and the tritium inventory in the repository will decay to insignificant levels in a few hundred years. Tritium can be incorporated into several different gaseous species, including hydrogen and methane. Depending on a range of factors, including site geology, these gases could in theory migrate to the ground surface at a relatively rapid rate if a gas pathway becomes established. Due to this reason, and the fact that there are substantial amounts of tritium present in the repository, it is necessary to consider it as one of the radionuclides capable of causing radiological exposures. However, note that for an appropriately-sited repository, it is expected that groundwater transit times will be sufficiently long so as not to cause radiological exposures to the general public from this radioactive gas (Thorne, 2005).

Radon-222
The long-lived radionuclide radium-226 is present in some waste streams, and it produces a continuous supply of radon as it decays. Radon-222 is the only significant gas produced by radioactive decay in ILW packages, but its short half life of only 3.82 days means that it is unlikely to migrate far from the production site. As radon-222 is a noble gas, it is not

expected to react with the waste package or its contents and can migrate away from its site of production by diffusive processes.

Carbon-14

Carbon-14 is created in reactor metals primarily as a result of neutron capture by nitrogen atoms through $^{14}N(n,p)^{14}C$ reaction during operation. C-14 is also present in graphite and organic wastes. C-14 atoms can be incorporated into several different chemical forms, both inorganic and organic, which will strongly affect its behaviour in the geosphere. While the retardation mechanisms of inorganic compounds such as carbon dioxide are rather well understood, the ways in which organic compounds migrate out of the repository are less clear. Organic compounds have low solubilities and are expected to be non-reactive in the near field. The main potential hazard from carbon-14 bearing gases produced in a repository environment, in terms of radiological dose, is thus considered to arise from organic species migrating from the repository, which, dependent on site specific features, could result in a radiological hazard to man.

Inorganic compounds considered here consist mainly of C-14 labelled carbon dioxide. The geochemical behaviour of CO_2 is strongly affected by the alkaline environment in which CO_2 reacts with materials in the repository near field to form inorganic calcium carbonate through a carbonation process. The release of C-14 from this source is then mainly controlled by the solubility and dissolution kinetics of calcium carbonate. The solubility of calcium carbonate in a solution saturated with portlandite is very low and natural analogues have been used to establish that a C-14 release from the repository would not be expected to occur (Dayal & Reardon, 1992). It is however possible that the repository conditions are very different from those of naturally occurring CO_2 storages and the repository CO_2 may be able to escape. It may also be possible for CO_2 to react with hydrogen to produce methane, although the carbonation reaction is expected to dominate over this reaction.

The main organic radioactive gas to be generated in the repository is expected to be methane. Methane has a low solubility and is likely to be non-reactive in the repository and in the geosphere. It can therefore be transported to the ground surface more easily than carbon dioxide, and, in the worst case scenario, may result in radiological doses to the public. Note, however, that many site specific geological features could act to retard the migration of gases generated in a repository environment, possibly for very considerable timescales such that any carbon-14 present will have decayed to insignificant levels (the half life of carbon-14 is of the order of 5000 years). Such effects of site geology need to be considered when the potential dose from repository-derived C-14 is considered in assessment studies.

GAS GENERATION PROCESSES

Gases, to which the above-mentioned radionuclides may become incorporated, are formed in the repository as a result of several processes, including the following:

Metal Corrosion

Corrosion reactions are the main source of the bulk gas, hydrogen, in the repository. Additionally, the release of C-14 and tritium from irradiated metals can result in the formation of several different C-14 and H-3 bearing gases, such as CO_2, methane and tritiated hydrogen. The rate of release of such gases is strongly dependent on the distribution of these atoms in the metal, but also on the availability of water and oxygen, and on temperature and pH, which all affect the metal corrosion rates. Additionally, the molecular form of C-14 in the metal affects the type of gas that can be formed. If carbon is present in the metal matrix as carbides, it is thought that hydrocarbons such as methane, acetylene, ethylene and ethane can be formed when the carbides become exposed on the surface and contact groundwater. If instead elemental carbon is present, the formation of organic gases is unlikely. The exact form of C-14 atoms in metals is the subject of on-going studies.

The inventories of different metals vary along with their corrosion rates, and these factors strongly affect the gas production rates. While the initial steady production of hydrogen is expected to be caused largely by the corrosion of Magnox, this inventory is expected to become consumed within approximately one hundred years of the repository closure (Hoch and Rodwell, 2003). On the other hand, steels will continue gas generation for tens, or even hundreds, of thousands of years.

Radiolysis
Dissociation of molecules by radiation can be a significant source of gas generation in an ILW repository at early times. Water within the waste package can be subjected to radiolysis from α, β and γ- radiation and produce hydrogen. If the water contains tritium, then the hydrogen produced will also be tritiated in the corresponding proportion. Radiolysis of organic compounds present in the wastes produces a variety of gases, of which hydrogen is expected to be the most important. If the organic wastes contain C-14 atoms, these may also become incorporated into the gas.

Microbial Degradation of Organic Wastes
The degradation of cellulose and small soluble organic molecules are considered to produce gases. Cellulose is initially hydrolysed to small organic molecules which are then degraded to produce CO_2 or CH_4. Methane is only produced in anaerobic conditions and in the absence of nitrate and sulphate ions. Nitrate and sulphate thus play a major role in preventing CH_4 production from the organic wastes.

Release of "Trapped" Radioactive Gases from Graphite
It is estimated that the inventory of C-14 in solid graphite accounts for about 80% of the total C-14 inventory (Norris & McKinney, 2008), but it is uncertain whether and at what rate the C-14 could be released. Some graphite may also contain tritium. The release of C-14 bearing gases and tritium can occur as graphite degrades, or by solid-state diffusion. Marsden *et al.* (2002) suggest that, in repository conditions, C-14 is leached at low rates from the graphite surface. Some authors (see, for example, Magnusson, 2002) note that both organic and inorganic forms may be formed from graphite, and that the releases are likely to be in the form of carbon dioxide and / or methane.

IMPACT OF REPOSITORY CONDITIONS ON GAS GENERATION RATES

It may be possible to affect gas generation rates by, for example, pre-treating waste or managing repository conditions during the operational phase. Gases generated in the repository during operation are, however, expected to be removed by ventilation and are not considered here.

Gas generation rates depend strongly on the waste inventories and package types used, but the potential for radioactive gas generation by a waste type is not always dependent on the amount of radionuclides in the waste. Graphite, for example, holds a large C-14 inventory, but its gas generation capability is expected to be small compared to that of metals and organic materials. In addition to waste inventories and package types, repository conditions also play an important role, as discussed below.

Availability of Water

The availability of water affects the corrosion rates of metals, thus having a direct effect on gas generation rates. Water is also required for microbial reactions; if there is little water available initially, the degradation of organic wastes during repository operation is restricted and increased amounts of gases could be produced after closure, when a repository could re-saturate with groundwater. Note that, after the onset of anaerobic conditions post-closure, the production of methane is favoured over the production of carbon dioxide.

pH

The interaction of groundwater with the cementitious backfill material is expected to cause the repository pH to rise to values of 12.5-13. In highly alkaline conditions the corrosion rates of metals and the solubility of radionuclides are considerably reduced.

Temperature

Temperature affects the corrosion rates of metals; increased temperature during the backfilling stage will cause a peak in the gas production rates. Temperature may also affect the repository pH and the solubility of repository materials. The solubility of portlandite, for example, decreases with increasing temperature. The repository pH is a result of complex interactions and may therefore not rise to values as high as calculated, thus affecting the gas generation rates. Additionally, temperature can have an impact on the microbial activity.

Microbial Populations

Microbial populations degrade organic material. If only low populations are present at the time of waste emplacement, more C-14 bearing organic material is left for consumption in the anaerobic rather than aerobic conditions. This means that an increased amount of methane could be generated after the repository closure.

Pockets of Anaerobicity

Pockets of anaerobicity may form in the waste packages before repository closure. This would increase the methanogenic microbial reactions and thus increase the generation rate of methane during operation.

Presence of Nitrate and Sulphate Ions

The presence of nitrate and sulphate ions increases the populations of nitrate-reducing and sulphate-reducing microbes, and slows down the growth of microbes involved in methane production. If, in particular, nitrate levels in the wastes are significantly reduced, the methane production is expected to increase (FitzGerald et al., 2004).

BARRIERS USED TO CONTROL GAS MIGRATION

Engineered and natural barriers in and around the repository act to prevent gases from reaching the biosphere, should a free gas phase form in the repository in the post-closure period. Several countries, such as the UK, France, Germany, Switzerland, Sweden, Finland, the US, Canada and Japan, have developed their preferred methods for underground disposal of radioactive wastes. Many of these involve disposal in a low permeability environment, where the host rock provides a major barrier to gas migration. However, in environments with significant groundwater flow, the role of engineered barriers in radionuclide retardation becomes far more important.

Engineered Barriers

The primary role of an Engineered Barrier System (EBS) is to contain short-lived radionuclides and to limit the long-term release of long-lived radionuclides. This can be done by controlling groundwater transport in and around the repository, by limiting radionuclide solubility, and by providing sorption surfaces for the radionuclides. An EBS consists of several possible elements. These include a suitable package in which the conditioned waste is placed, a suitable encapsulant, an appropriate backfill material, and effective sealing of the repository from the surface environment.

Waste packages

The first barrier repository gases encounter is the waste package itself and the encapsulant material used. The waste package typically comprises a steel or concrete container, within which the waste is immobilised using an immobilisation matrix such as cement grout. The main function of the waste package is to contain short-lived radionuclides until they have decayed to insignificant levels. The waste container and immobilised wasteform provide a barrier for groundwater access to the wastes, thus limiting the dissolution and transport of radionuclides with the groundwater. Containers may be vented in order to prevent the build-up of internal gases and therefore some long-lived nuclides are expected to be released with gases escaping from the waste containers.

Wastes must be converted into passively safe, solid wasteforms, which means immobilising liquids, sludges and fragmented solids. A common immobilisation matrix is cement. Cementitious materials condition the chemical environment of the wasteform to high pH values, which ensures the low solubility of many radionuclides. Additionally, a high pH environment will decrease the corrosion rates of steel containers and thus reduce the release rate of radionuclides.

Backfilling and sealing of the repository

After all the wastes have been emplaced in the repository, backfilling, sealing and closure is possible. A role of the backfill may be to limit groundwater flow around the wastes and to create favourable physical and chemical conditions. Lu and Conca (2003) reported that backfills are often divided into two categories: chemical backfills such as cement, iron, phosphates and MgO, and hydrological/physical backfills such as clay, salt and cement.

For the assessment studies to date of a UK repository concept, the backfill material for ILW has been assumed to consist of a cementitious material, which will act as a chemical buffer and create uniform alkaline and chemically reducing conditions. Backfill will also provide a surface for radionuclide sorption due to its relatively high porosity. Additionally, it can be used as mechanical support and as a barrier to control groundwater flow around the wastes. Another commonly used backfill material is bentonite clay. This swells on contact with water and plugs all openings, thus ensuring diffusion-controlled solute transport.

The Swedish and Finnish repository concepts use crushed rock as a backfill, with the purpose of forming a hydraulic cage around the wastes. The hydraulic cage relies on the highly permeable layer of crushed rock to form a preferential pathway for groundwater flow and thus prevent advective flow driven by hydraulic gradients through the wastes. This should result in the diffusion-controlled transport of radionuclides. The long-term behaviour of the hydraulic cage is, however, not known and the crushed rock zone could provide a preferential pathway to gases, should a release occur.

The use of magnesium oxide forms an important part of the disposal concept for the WIPP repository in the US. Sacks of MgO are to be emplaced with the waste, and the wastes will be encapsulated in the repository by creep closure of the host rock (a bedded salt) which will form a low-permeability barrier around the wastes. The MgO sacks will then break open and consume moisture and CO_2 that may be produced in the repository. It will also buffer the pH such that actinide solubilities are limited. About 1,000 tonnes of MgO will be placed in the repository (DOE, 2004).

After backfilling the repository, repository disposal vault entrances and exits will be sealed using low-permeability materials. Seals can also be placed at other key locations to prevent the underground tunnels, drifts and shafts from acting as preferential pathways for groundwater flow and radionuclide transport. The migration of gases into the host rock can also be controlled; for example, by shotcreting the vault walls or grouting the most transmissive features. This may affect the rate at which gas is generated in the vault (e.g. by limiting water availability) and the rate at which generated gas then migrates out of the vaults.

REPOSITORY DESIGNS

Repository designs need to deal with complex issues, such as the containment of radioactive gases, temperature increases, and isolation of waste types which may release molecules that affect the solubility or sorption of radionuclides in the waste. After encountering the engineered barriers in the repository, gas migration will depend on the properties of the host rock and the surrounding environment. Low permeability rock surrounding the repository ensures low rates of groundwater flow and thus limits radionuclide dissolution and transport with the groundwater. Also, radioactive decay, dispersion and sorption in the host rock will limit the radionuclide migration. Several disposal concepts have been proposed internationally to deal with these issues. Hicks *et al.* (2008) divided these into four groups as summarised in TABLE 1.

SUMMARY

An ILW repository may contain a significant inventory of C-14, H-3 and Rn-222. These radionuclides may form radioactive gases that can migrate from the repository to the biosphere with the bulk gas. Due to the short half-life of Rn-222, its radiological impacts are considered negligible. Tritium has also got a relatively short half-life, and so most research concentrates on the potential release of C-14 from the repository.

Corrosion of metals in the repository produces large amounts of non-radioactive hydrogen gas, but, if metals or the surrounding water contain C-14 or H-3 atoms, these can also be released in gaseous form. The form of C-14 atoms in the metal structure plays an important role: if C-14 is present as carbides it can be hydrolysed to form methane or other hydrocarbons; if C-14 is present in its elemental form, it is unlikely to form a gas. Another possible source of radioactive gas is graphite, as it holds a substantial inventory of C-14 which may form methane in repository conditions. Understanding the release behaviour of C-14 from graphite is currently a key concern for repository safety.

TABLE 1: Common features of repositories in different geological settings. (Based on Hicks *et al.*, 2008).

Disposal in:	Weak rock with little or no groundwater flow	Strong rock with little or no groundwater flow	Strong rock with significant groundwater flow	Plastic evaporite rock with no groundwater flow
Countries	Belgium, France, Japan, Switzerland	Canada, Japan	Sweden, Finland	U.S.
Host rock	Indurated/plastic low permeability sedimentary rock	Crystalline rock	Crystalline rock, carbonate	Evaporites: salt dome, bedded salt
Primary barrier	Host rock most important barrier. Self-healing of cracks and fissures an important property of the rock.	Low-permeability environment most important barrier. Importance of EBS increased if there is groundwater flow.	Role of EBS far more important than in other concepts. Cementitious materials limit groundwater flow and provide long-lasting alkaline environment.	Low permeability salt rock encloses waste as result of salt creep.
Secondary barriers	Alkaline and reducing environment due to cementitious materials will limit radionuclide solubility and increase sorption.	Alkaline and reducing environment due to cementitious materials will limit radionuclide solubility and increase sorption.	Host rock with low hydraulic conductivity.	Engineered low permeability seals will prevent shafts from becoming preferential pathways.
Waste packages	Concrete (sometimes bitumen) used as immobilisation matrix. Emplacement packages proposed in several concepts.	Several immobilisation matrices proposed. Cement often used to provide alkaline conditions.	Cement conditioned wastes placed in metal containers or concrete packages.	Steel containers used for handling rather than to perform a specific role in radionuclide retention.
Backfill	Grout backfill sometimes used. Sealing of disposal tunnels, caverns, etc. important in preventing these from becoming preferential pathways.	Rock bolts and grout (sometimes bentonite) backfill often used for mechanical stability.	Crushed rock used to divert flow away from waste packages.	MgO used as buffer to limit actinide solubilities.

Radiological waste repositories can be located in various geological settings. The properties of the surrounding rock determine the importance of engineered barriers to long term safety. If the repository is located in a low permeability host rock where there is little or no groundwater flow, the host rock can be regarded as the most important factor in radionuclide retardation. However, when groundwater flow rates are increased, engineered barriers play a much more important role.

Acknowledgements
The authors would like to thank Ove Arup & Partners Ltd and EPSRC for their financial support, and the NDA for their assistance with the project.

References

Dayal, R. & Reardon, E. (1992). Cement-based engineered barriers for carbon-14 isolation. *Waste Management.* 12, 189-200.

DOE (2004). 2004 WIPP compliance recertification application – appendices. DOE/WIPP 04-3231.

FitzGerald, P. L., Rees, J. H. & Sanchez-Friera, P. (2004). Sensitivity of carbon-14 gas production in a deep repository to variations in repository conditions. Report for UK Nirex Ltd. SERCO/ERRA-0460.

Hicks, T. W., Baldwin, T. D., Hooker, P. J., Richardson, P. J., Chapman, N. A., McKinley, I. G. & Neall, F. B. (2008). Concepts for the geological disposal of intermediate-level radioactive waste. Report for NDA. 0736-1 Version 1.1.

Hoch, A. R. & Rodwell, W. R. (2003). Gas generation calculations for generic documents update. Report for UK Nirex Ltd. SA/ENV-0514 Version 2.

Lu, N. & Conca, J. L. (2003). Plutonium behaviour in brines after equilibration with periclase (MgO) backfill. *Acta Mineralogica-Petrographica.* Abstract Series 1, p. 65. Szeged.

Magnusson, Å. (2002). Measurement of the distribution of organic and inorganic ^{14}C in a graphite reflector from a Swedish nuclear reactor. Department of Physics, Lund University, Sweden.

Marsden, B. J., Hopkinson, K. L. & Wickham, A. J. (2002). The chemical form of carbon-14 within graphite. Report for UK Nirex Ltd. SA/RJCB/RD03612001/R01 Issue 4.

Norris, S. & McKinney, J. (2008). UK graphite wastes: issues relating to long-term management and disposal. Presentation to 7th EPRI International Decommissioning and Radioactive Waste Workshop, 29th October 2008, Lyon.

Thorne, M. C. (2005). Development of increased understanding of potential radiological impacts of radioactive gases from a deep geological repository: post-closure significance of H-3. Report for UK Nirex Ltd. MTA/P0011b/2005-9 Issue 2.

Author Index

Subject Index (by Keywords)

Breinigsville, PA USA
23 July 2010
242343BV00006B/10/P

9 781847 559135